CURRICULAR MANAGEMENT
and socioformative approach

CURRICULAR MANAGEMENT

and socioformative approach

Sergio Tobon

APA sixth edition:

Tobon, S. (2016). *Curricular management and socioformative approach.* Lake Mary, USA: Kresearch.
246 p.

ISBN 978-1-945721-16-8
Education –curriculum. L. T.

Ekap university
801 International Parkway • 5th Floor
Lake Mary
Florida
32746
USA
Tel. 407-830-1200

This book was translated by
ROSALIA NALLELI PÉREZ ESTRADA
rosalia_na@hotmail.com

http://kresearch.us/
contact@kresearch.us

Online shop
www.amazon.com

Member of the National Chamber of
the Editorial Industry

First Edition, December, Lake Mary (USA), 2016
ISBN: 978-1-945721-16-8
DOI: http://dx.doi.org/10.24944/isbn.978-1-945721-16-8

Printed in USA

Table of contents

PRESENTATION

The purpose of this book is to give you some practical ideas which may help you to implement the curriculum, according to the nowadays social challenges and the future tendencies. It has as a base the management of the learning quality and the formation of people with critical thinking, entrepreneurship and critical behavior. To achieve this, basic conceptual and methodological aspects of competencies are analyzed. Besides, planning, execution and evaluation of curriculum are specified from the institutional direction to its implementation with students through the methodology of formative projects. This can be achieved based on flexibility, transversality, collaboration, participation of the educative community and the continuous improvement of teaching practices.

Formative projects emphasize on learning the established competencies in a determined graduation profile, through the action of projects with the students, on the articulation of knowledge frame and by considering different contexts: personal, familiar, social, cultural, laboral- professional environmental-ecological, recreative-sportive, artistic, disciplinary and research contexts.

A project consists on a set of articulated activities, with the objective of identifying, interpreting, arguing and solving a contextual problem, all based on collaborative work, the service and the application of complex thinking. In this form the traditional subjects, based on content, are proposed to be moved on projects; which contribute to transform reality, besides making the integral formation easier.

This book is based on different experiences of curricular management by competencies, implemented from a flexible, integral and humanistic perspective by researchers in education and management of human talent. Such experiences have been done in most of Latin American countries as well as in some european countries (such as France and Sweden). From those experiences, the traditional approaches of competencies, centered on tasks and the fragmentation of performance are overcome, as it is the case of the functionalist, behavioral and traditional constructivist approaches. On their place, alternative and more suitable approaches for the knowledge society are emerging as it is the case of the socioformative approach. This is centered on the integral development before contextual problems, the development of complex thinking and the entrepreneur spirit.

The construction and application of the socioformative approach in education and in the organizations, have been done since the 90´s. Nowadays, this approach is followed by multiple researchers in several countries and from various points of emphasis. This line of work follows the referents of complex thinking, the fifth discipline, the development of human scale and the learning how to learn and how to undertake. According to this, the socioformative approach emphasizes to form and to strengthen the ethical project of life and the competencies, with the goal of achieving personal self-fulfillment, the construction and reinforcement of social tissue, recreation, the continuous searching of economical entrepreneurial development and the equilibrium on ecological suitability.

In spite of the fact that nowadays there is a great quantity of educative institutions which follow the referents of socioformative approach, this is not over yet. This is a line of work in permanent building, in which the theoretical and methodological proposals have been generated, validated in the educative practice and in the management of human talent. However, there are still many questions to be answered. This book is an invitation to continue strengthening the building of the socioformative approach, to have as a base different lines and schools of work, study and research, so that this may give way to a better quality management of learning processes in the educative institutions.

In the book, it is not the intention of presenting all the referents of the socioformative approach but to share some basic guidelines concerning to basic processes implied in the curricular management and to ensure the transformation of education within the framework of the challenges of the knowledge society. In a more concrete manner some aspects, as the following ones, are mentioned here:

1. The socioformative concept of competencies by applying the conceptual cartography, which is a strategy for construction of scientific concepts.
2. The study of the context from problems and with a prospective analysis.
3. The elaboration of graduation and entry profile by competencies, based on the study of contextual problems.
4. The design of curricular maps (nets) with integrality, flexibility and systemic approach.
5. The competencies that teachers and principals must possess to make the integral formation real and to make competent students.
6. The synthetic methodology of formative projects to implement the curriculum.
7. The mediation of the teacher, based on the ten minimal actions that socioformation proposes to ensure the educative change.

Much like in other books of the author, in this one, it is insisted to change traditional, educative practices through the educative, action research which must be done by principals and teachers as part of their job. In the action research, non -relevant educative activities are found, according to the challenges of contexts; it is thought about them. There is a practice of theorizing and they are changed through collaboration, systematization and publishing of achievements to motivate teachers towards the pedagogical change. All of this is done through four principal axis: addressing, planning, acting and evaluating. The methodology of curricular management proposed in this book follows these referents.

It is intended that the curriculum management can be given from the frame of a practice with discourse. It means; based on principles and concepts of pedagogy, social sciences and complex thinking. This is essential so that curriculum management can be based on solid argumentations and in a systematic methodology, something that with frequency is absent in this field. This is why you are invited to do new research and applications which lead to a progressive consolidation of new paradigms of competencies, as a key proposal for the improvement of quality of education in Latin America.

The book is directed, firstly, to people comissioned of managing and applying curriculum of diverse educative institutions, such as principals, coordinators, directors of projects of educative innovation, people who work in offices of institutional auto evaluation and teachers. Secondly, it is directed to researchers and agencies of accreditation and evaluation in quality of education, because it offers a model about criteria and basic processes which must be considered in the curriculum management, from the academic quality approach.

CHAPTER 1

Conceptual cartography of the competencies

How to overcome the fragmentation of the performance and to center it in the integral performance of people?

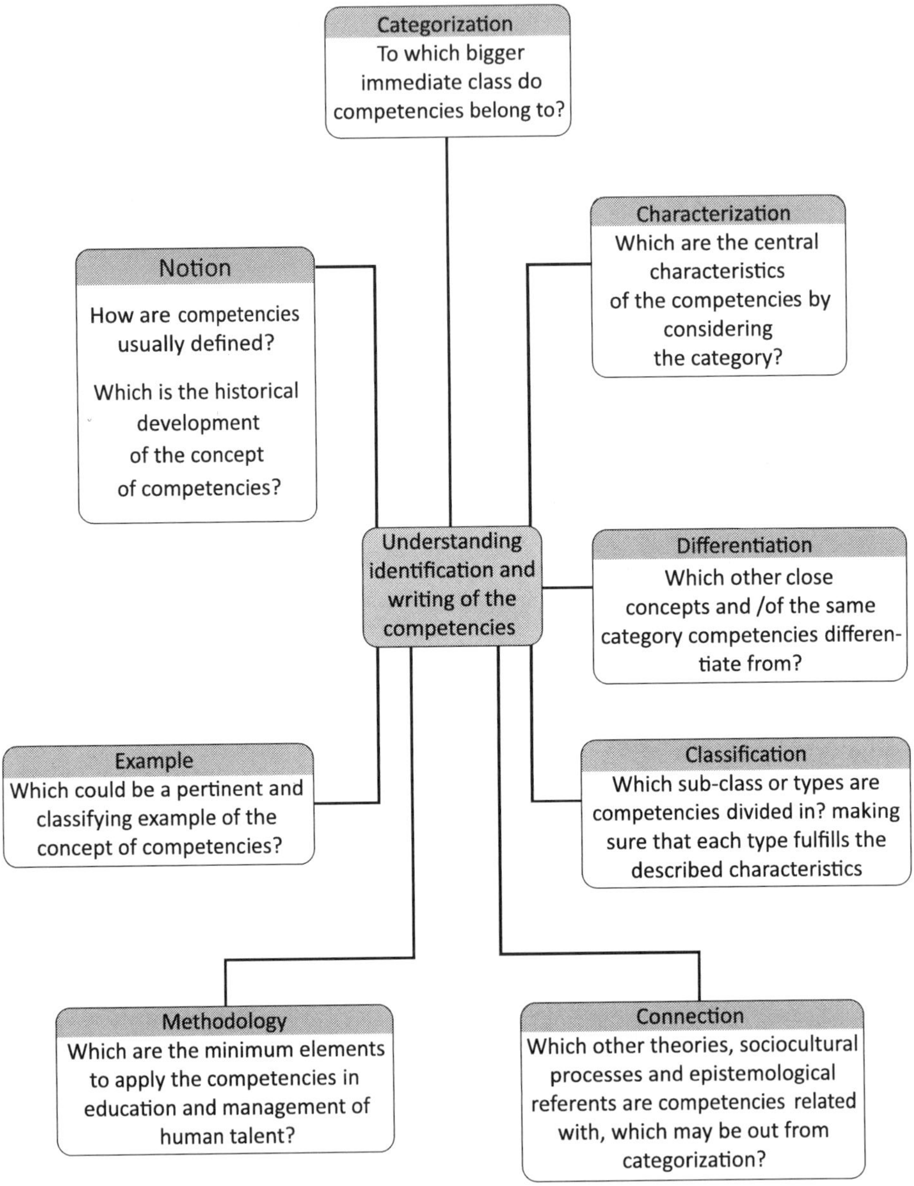

Figure 1.1 Basic questions to analyze the concept of competencies.

METHODOLOGY OF THE CONCEPTUAL CARTOGRAPHY

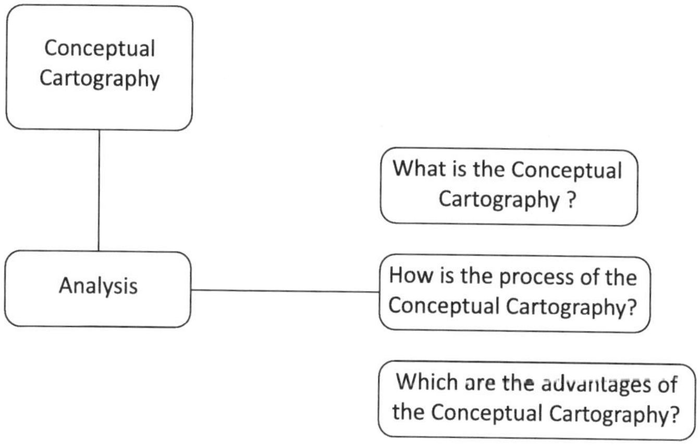

What is the conceptual cartography?

The conceptual cartography is a strategy adapted by Tobón (2004, 2013 b) from the conceptual minded- facts and the mental maps. It consists on orienting the analysis, the building and the communication of a concept or theory through a series of key questions which are responded through the bibliographic analysis, the collaborative work and the search of agreements towards the fundamentals. This is done in a graphic map (in the form of mental map) which is complemented with the writing of an argumentative of the answers to the questions and its relations.

In the figure 1.1 the taken questions to build the concept of competencies are described in this chapter. Eight questions are generally developed (Tobón 2013a,b) but in this chapter, one more question has been added, which is question number three, referred to the historical development of the concept.

How is the process of the conceptual cartography?

The process of the work with the cartography implies to review, first, the previous knowledge concerning to the concept or theory by trying to answer the given questions. From this on, a collaborative work of research analysis is begun in the proposed order. Once analyzed and answered the questions, the next step is to make the map of the conceptual cartography to indicate the concept in the center; located next to an image which represents it and which helps to understand its meaning. Then each one of the axis or questions is spread out.

Next, the most relevant aspects of each axis are indicated with words or key words, and there is an establishment of relationships between a single axis and among axes and the other axes. This process is accompanied by images, logos and prominences which help to understand the described elements. When one relationship is finished its sense is indicated with a key word or a phrase. At last, the map is explained with all its main aspects; it is socialized and the questions are reviewed again until achieving solid answers.

Which are the advantages of the Conceptual Cartography?

The conceptual cartography has the following advantages; regarding to other similar strategies such as conceptual maps and mental maps.

1. It tackles the key main ideas of every academic concept or theory such as in its construction as in its learning and communication. These main ideas have been always used in science but implicitly. The fact of expliciting them strengthens the understanding and appropriation of knowledge to resolve problems of the context.
2. It articulates as individual work as collaborative work in the building and learning of concepts and theories which deepens the process.
3. It tackles new knowledge from previous knowledge of people and that makes learning easy.
4. It helps to deconstruct preconcepts based on the review of external sources, the continuous questioning and establishment of agreements concerning key elements.
5. It promotes reading understanding and critical lecture through key questions about different sources of information.
6. It prevents the copy of the texts, very common nowadays due to the fact that people have the need of building texts to argue their answers and their relationship to each one of the questions.
7. It develops the creativity because it invites to formulate new

questions which have a major development or deepness of the concept. For example, these questions may be asked: How could we innovate the concept? Which are the problems or emptiness of the concept? Which is the possible future of the concept? etc.

EXEMPLIFICATION OF THE COMPETENCIES

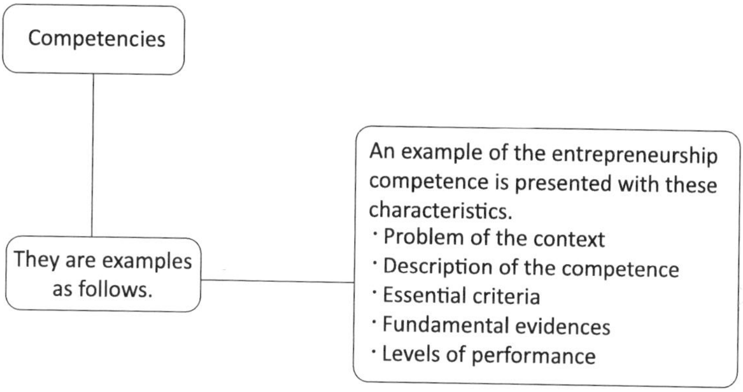

To understand the competencies on chart 1.1. An example of entrepreneurship competence is presented. This competence is an integral performance that people develop to create, to begin and to make a project to work out, so that it contributes to solve a need (or problem). To orient the formation and evaluation of this competence there are, among others, five minimum criteria which must be verified in three concrete essential products (evidences).

Chart 1.1 Example of competency from a Socioformative approach.

Institution: Autonomus National University	
Title of the competence: Enterpreneurship.	
Version: 3.0(June 2012)	Autor: Jorge López Campusano.
Revision: July 30th , 2012	Copy editor: Adriana Marin López Carla Giraldo Campos.
Contextual problems which try to tackle the problem	
· How to create social or company projects which resolve the problems systematically? · How to achieve projects to continue with the time and to face difficulties with appropriate strategies? · How to join the business process with the ethical project of life of people and to make sure about success and pertinence, personally and socially?	

Description of the competence	
To manage creative and innovative projects which contribute to identify, to set and to resolve the social and productive problems based on systemic methodologies and challenge of the context.	
Criteria (patterns to value the competencies).	*Evidences (test to verify the criteria).*
1. To do entrepreneurial projects on the frame of the ethical project of life. 2. To propose the idea of a feasible project to solve a need of the context, according to a determined diagnostic. 3. To plan an entrepreneurial project according to the needs of the context and make a comparison with other similar projects. 4. To execute projects of entrepreneurship by considering the difficulties as opportunities of improvement and innovation, with economic and social viability. 5. To argue the fundamental components of a project of entrepreneurship and the methodologies to undertake, according to a determined market niche and the socioeconomic context.	· Register of the interview · Information of a process of entrepreneurship done. · Show products and/or offered services.

With the purpose of complementing the example, a learning map is shown on the frame 1.2. To orient the formation and evaluation of the entrepreneurial project as it is done every day in a program of medicine. This learning map is an adaption of the rubrics of work by competencies and it is composed of levels of performance: pre-formal, receptive, decisive, autonomous and strategic. Based on these levels, the progress of the people on their formation is established, and actions of help are also complemented by taking into account the evaluation (self-evaluation), the co-evaluation (peer evaluation), heteroevaluation (evaluation made by a mediator or an expert) and the metaevaluation (evaluation of the self-evaluation process to improve it and to search it to be more pertinent each time).

Frame 1.2. Learning map of the entrepreneurship competence

Program of studies: medicine
Project: Fulfillment of a process of entrepreneurship to satisfy the need of the context.
Time: 2 years **Evidences:** Register and observation of an entrepreneurship process

Criterion	*Preformal level*	*Receptive level*	*Resolutive level*	*Autonomous level*	*Strategic level*
1. To make entrepreneurial projects within the framework of an ethical project of life.	To make entrepreneurial projects within the framework of an ethical project of life.	To make entrepreneurial projects with some personal motivation besides answering external circumstances.	To get challenges towards entrepreneurship and to establish them with clear goals. To make projects of entrepreneurship based on the vital needs of growth and to find the self realization in these projects.	To make the entrepreneurial projects by meeting the universal values: Responsibility Respect Collaboration Justice Honor	To show a constant perseverance in the process of entrepreneurship and to assume the difficulties as opportunities. To have a mission and a vision as entrepreneur. The person is conscious of the difficulties but shows strength to face them and to get over them.
Weighting	5 %	10 %	15 %	22 %	25 %

2. To propose the idea of a project which is feasible to solve a need of the context, according to a determined diagnosis.	To propose the idea of a project to solve a need of the context without a clear argument.	To propose the idea of a project to solve a need of the context, identified through a diagnosis.	To propose the idea of a feasible project to solve a need of the context, by giving something towards the existent solutions.	To propose the idea of a project, that concerns a need supported with clarity based on the pertinent diagnosis. The need of the context is perfectibly limited and the project possess the necessary support	To propose a creative and innovative idea of a project before the existent solutions.
Weighting	5%	10%	20%	25 %	25 %
3. To plan an entrepreneurial project according to the needs of the context and to make a comparison with other similar projects.	To plan in a general form a project without clarity in the administrative part or in the strategies to keep it on time.	To plan an entrepreneuria l project.	To plan an entrepreneurial project by following some formal, determined criteria. The project responds to an established need in the diagnosis.	To plan an entrepreneurial project with a clear support of contributions	To plan an entrepreneurial project with creativity and innovation which contains strategies to keep it in the time and to prevent failure.
Weighting	5%	15%	20 %	30%	20%

Program of studies: medicine
Project: Realization of an entrepreneurial project to satisfy a need of the context.
Time: 2 years **Evidences:** Register and observation of an entrepreneurship process

Criterion	*Pre formal level*	*Receptive level*	*Decisive level*	*Autonomous level*	*Strategic level*
4. To execute entrepreneurial projects by approaching the difficulties of improvement and innovation with economic and social viability.	To execute the projects of entrepreneurs hip without changes in their setting	To execute the entrepreneuria l projects by identifying the difficulties and by analyzing the improvement options.	To execute the entrepreneurial projects by improving in the activities, goals and resources with the objective of ensuring the achievement of established goals.	To execute projects of entrepreneurship with autonomy and initiative by implementing the strategies, to face the problems which may appear.	To execute entrepreneurial projects with strategies which prevent possible problems and with a high degree of perseverance, to achieve the established goals. To generate social and financial profitability. To implement innovations from tackling the problems, based on contextual opportunities.
Weighting	1 %	2 %	5 %	15%	25 %

5. To argue the fundamental components of an entrepreneurial project and the methodology to undertake according to a determined market niche and the socioeconomic context.	To have a general knowledge of an entrepreneurial project and of the methodologies to undertake, according to the required need.	To have structured concepts of different strategies of entrepreneurship, based on experiences in the area.	To argue the best strategies conceptually to undertake, according to the theoretical references.	To argue and to give examples from a socioeconomic context, the most pertinent strategies to undertake, by considering a determined need to be solved.	To expose with clarity in different contexts the entrepreneurial project and to achieve social and financial help for this.
Weighting	1%	2%	3%	5%	5%
Process of evaluation	**Achievements**		**Recommendations**		**Level achieved**
Self-evaluation					
Co-evaluation					
Hetero-evaluation					
Meta-evaluation					

NOTION OF the Competencies

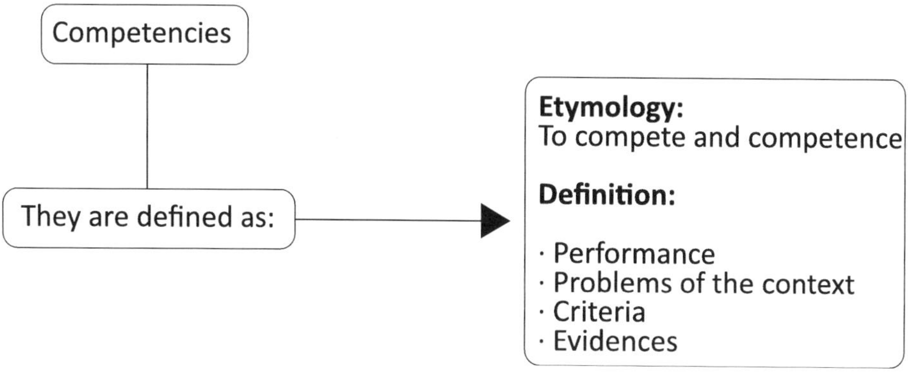

Which is the etymology of the Word Competencies?

This question must be answered when an educational project with this approach is done, because some programs of formation in education are frequently established without a deep understanding of what competencies are. It can be done in the macro curricular area or in a concrete formative space (subject, module, formative project, etc.) and it makes that expected results do not get achieved. For example, in a conference given by the author of this book in Portugal, a professor stated that the best form of tackling competencies was to make students to compete for the best grades through contests. This leads us to the following question: competencies mean competition?

We have to begin by saying that in spanish, competencies have two different meaningful lines completely different and opposed, according to the dictionary of the Royal Spanish Academy (RAE, 2012): to compete and to compete with . To compete with refers to comparison, rivalry and to stand out in relation to others, and it gives way to meanings such as competition, competitively, competitive, etc. This comes from the greek terms: agon and agonists which refer to compete among people, as it is in the competition of the Olympic Games.

To compete, for its part, refers to incumbent upon or suitability, (RAE, 2012). This is what corresponds to each person with management and responsibility. For example, when a technician in tourism competes to create and formulate touristic projects for a certain region. This conception relates, for its part, with the Greek term ikanotis which has to do with suitability of people and not of competition.

In education, and in the management of human talent, we use the term "competencies" with this second definition. If we have this clear, it will help to avoid unfair critics to the concept of competencies a lot, which is related to competitiveness, when its core point is the emphasis on developing with suitability and responsibility in a collaborative scope. Therefore, it is the moment to use this concept in education and to start transforming the formative practices.

How are competencies defined?

There are multiple definitions about competencies nowadays. This is the result of several perspectives and proposals of implementing the concept as much in the educational as in the organizational world. On frame 1.3, there is a list of some definitions as well as a criticism of them resulting from the experiences that we have had, concerning to its implementation in education, on the frame of complex thinking.

According to these definitions, it can be said that competencies are performances before situations of the context. To achieve this, they articulate several types of sets of knowledge, such as knowledge, processes of understanding, metacognition, values, attitudes, abilities and skills to search for making an impact on the environment through determined results.

This means that there is not competence without application. For example, a student can know many things about research, motivation and several abilities on this field (ability of looking into information, ability to synthesize, ability to systematize the information, etc.) but if he does not present relevant products of research according to the educative cycle (such as exploratory studies, speeches, articles, etc.) therefore, he does not have the research competence.

Frame 1.3 Some traditional definitions of the concept of competencies.

Traditional and Current definitions	*Criticism*	*What does the socioformative approach suggest?*
Spencer and Spencer (1993) consider that the competence is "an underlying characteristic of the individual which is by chance related with an effective or superior efficiency in the work or in a situation, defined in terms of a criterion".	This idea was very relevant some years ago to start setting the competencies model. Nowadays we understand that competences are not underlying nor are they external, but a combination of both attributes. The current concept of competencies considers the average efficiency and not only the superior efficiency.	From a socioformative approach, what exists is the acting process, and this has as underlying internal aspects as external aspects, articulated in web to achieve determined goals.
"Repertoire of behaviors (they integrate aptitudes, features of personality and knowledge) that some people dominate better than others, which makes them efficient in a determined situation" (Levy-Leboyer, 1997).	The behavior is a part of the competencies, and not all the competence, even when it is a repertoire. The competencies must not be exclusively what some people dominate better than others, because comparisons among people must transcend.	The socioformative approach is not based on the comparison of successful people with unsuccessful people in a situation. It rather seeks that people compare with themselves from suitability criteria.

Schmelckes, dated by Barron (2000) understands as competence "a complex, which implies and includes, in each case, at least four components: information, knowledge, (which refers to appropriation, processing, and application of the information) ability and attitude or value" page 31)	It is relevant to set out the term complex in the sense of diverse parts interrelated. This definition lacks the reference of quality-suitability, an essential component of the competencies.	The competencies are a complex acting process, in the sense that they articulate and interweave diverse, essential components in the performance with pertinence. More than information and knowledge, knowledge of systemic processes are articulated in competences.
Competencies are understood as an efficient performance in determined situations, which help in the acquired knowledge and in other cognitive resources (Condemarin and Medina 2000)	Differently from the previous definition, in this definition the reference to the term quality can be seen, through the term efficient. This definition tends to show an image of the competencies as internal cognitive processes when they are also compound of external factors, and of relation with the context.	The competencies are processes that articulate systematically internal elements with external elements in the behavior in the frame of dynamic relations with the environment.
Competencies are "a suitable performance which emerges in concrete tasks in a meaningful context. It is a concept used with property which acts to be applied in a determined situation, in an enough, flexible manner, as to give several and pertinent solutions (...)(Bogoya, 2000.p 11...)	This is a definition that has been followed a lot in Colombia. As main achievements, there is the reference to performance with suitability, the context, flexibility and several solutions that stand out. It does not make reference in an enough explicit form to the integration of the knowledge.	In the socioformative approach of competencies, besides Bogoya´s proposal, it emphasizes the integration of knowledge in ethics and in the metacognitive processes.
A competence is a capacity for the development of relatively new tasks, in the sense that they are different to the routine tasks made in class or which were set in different contexts from those that were taught. (Vasco, 2003, page 37).	Competencies are not capacities, but an integration of the performance of capabilities, attitudes, values and meta-cognitive processes around concrete goals. There are not always new goals in competencies, because they can be applied in known, predictable and not new activities .	The Socio-Formative approach states that competencies are framed as in known activities as in new activities.
Massot and Feisthammel (2003) define competencies as an organized set of structures of behavior that respond to a real background in the global professional framework.	Behaviors are only a part of the competencies and they are not their core.	Competencies are structures of complex and dynamic performance which integrate behaviors, attitudes and knowledge in the frame of systemic relations of the person with the contexts in which this person develops.

It is necessary to complement this first approach to competencies with the different approaches of competencies that exist nowadays. Basically, we have four key approaches in Latin America: behaviorist, functionalist, constructivist, and socioformative approaches. In frame 1.4 these four approaches are described as well as their emphasis in the creation of competencies.

Frame 1.4 Conception of competencies in the different approaches. (Tobón, 2012a).

1.- Behavioral approach	It emphasizes in assuming competencies as key behaviors of people for competitiveness of organizations.
2.- Functionalist approach	It emphasizes in assuming competencies as sets of attributes that people must have to fulfill with the goals of professional labour processes framed in defined functions.
3.- Constructivist approach	It emphasizes in assuming competencies as processes which are composed of abilities, knowledge, and attitudes to solve difficulties and dysfunctions in social, professional labor processes from an organizational framework.
4.- Socioformative approach	It emphasizes in taking over competencies as integral performance to solve activities and problems of the context with suitability, continuous improvement and ethics.

What are competencies from a socioformative approach?

From socioformation, the concept of competencies has been built based on the proposals of complex thinking by considering the formation of people with critical thinking, creativity, leadership, collaboration, responsibility, respect and social and business entrepreneurship. According to this, competencies are defined as:

Integral performances to identify, to interpret, to argue, and to solve problems of diverse contexts, with feasibility, continuous improvement and ethical behavior, by developing and setting into action in an articulated manner to know how to be, to know how to coexist, to know how to do and to know how to know, with the goal of promoting self-realization, building and holding on social tissue: The continuous search of economic-entrepreneurial development, recreation and enjoying of life, the artistic creation, research, care and protection of environment and of alive species (Tobón, 2013 a)

There are different definitions of competencies, but there is still a lack of emphasis in solving problems, in ethics and continuous improvement to achieve a change of traditional, educative practices and to form for the knowledge society, which is emerging around the world. Socioformation centers in it and sets it into action based on didactic methodologies focused on projects and on collaborative work.

In figures 1.2, 1.3 and 1.4 details of socioformative perspective of competencies are illustrated.

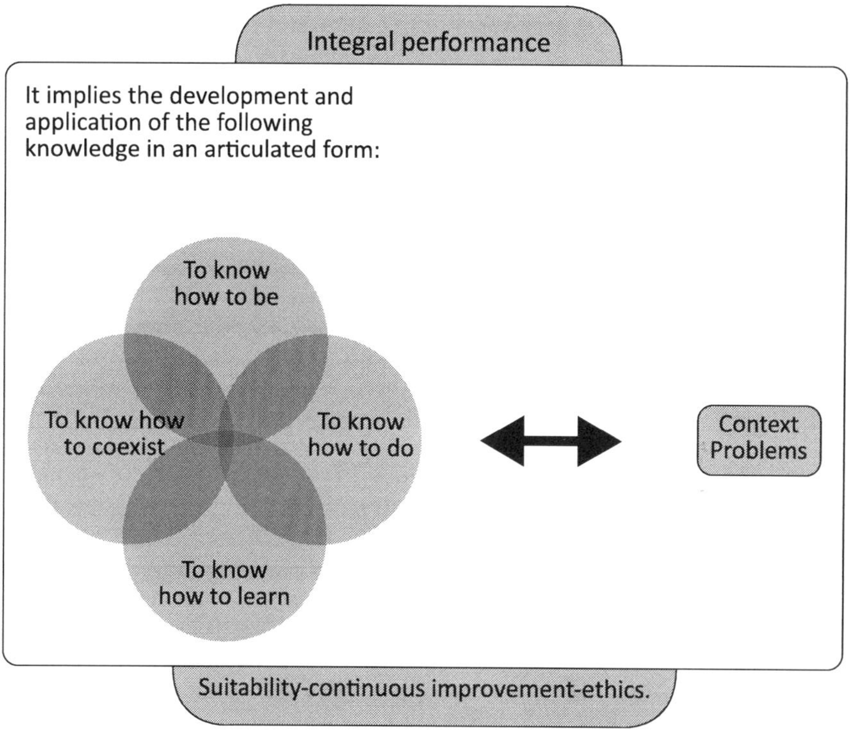

Figure 1.2 Central concept of definition of competencies from Socioformation.

Figure 1.3

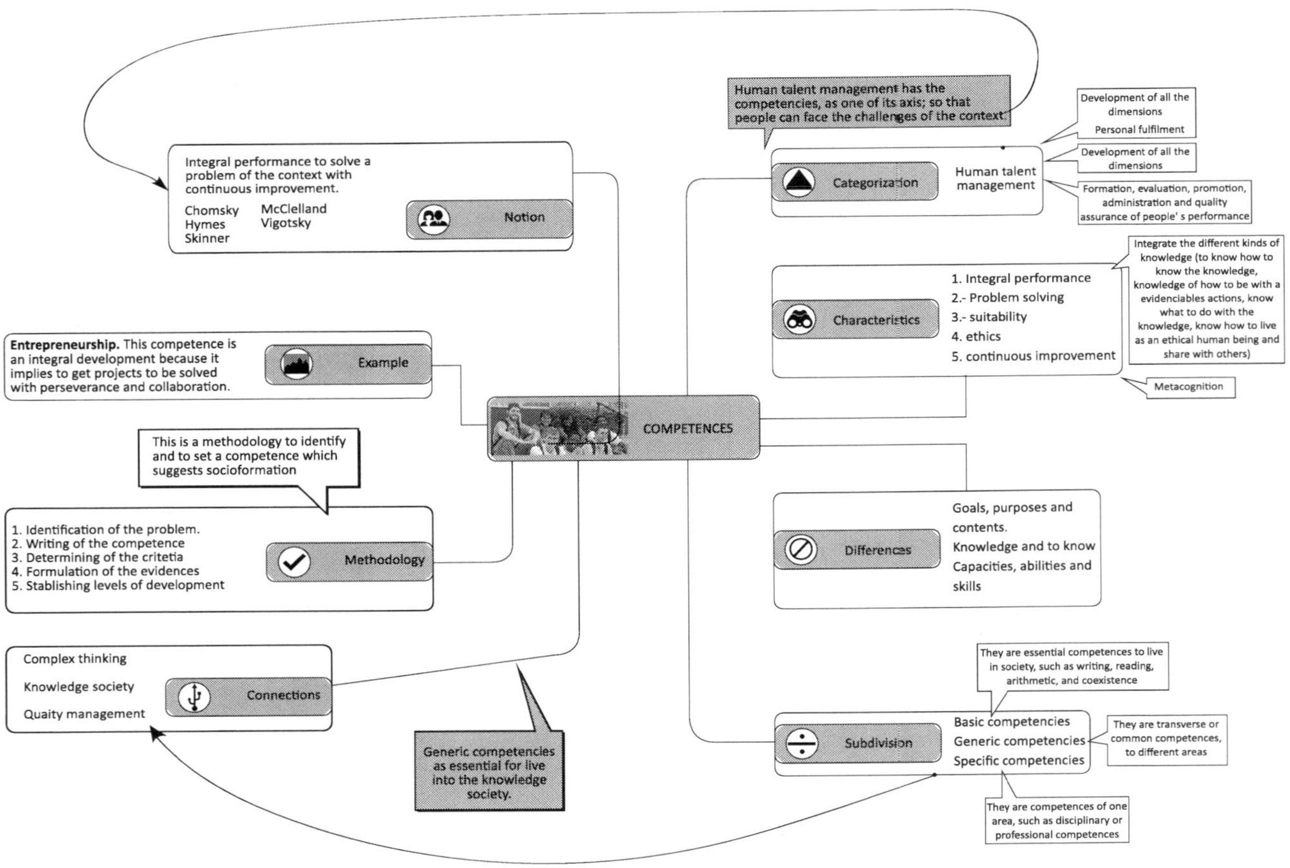
COMPETENCES
Notion
Integral performance to solve a problem of the context with continuous improvement.
Chomsky
Hymes
Skinner
McClelland
Vigotsky
Example
Entrepreneurship. This competence is an integral development because it implies to get projects to be solved with perseverance and collaboration.
This is a methodology to identify and to set a competence which suggests socioformation
Methodology
1. Identification of the problem.
2. Writing of the competence
3. Determining of the critetia
4. Formulation of the evidences
5. Stablishing levels of development
Connections
Complex thinking
Knowledge society
Quaity management
Generic competencies as essential for live into the knowledge society.
Human talent management has the competencies, as one of its axis; so that people can face the challenges of the context.
Categorization
Human talent management
Development of all the dimensions
Personal fulfilment
Development of all the dimensions
Formation, evaluation, promotion, administration and quality assurance of people' s performance
Characteristics
1. Integral performance
2.- Problem solving
3.- suitability
4. ethics
5. continuous improvement
Integrate the different kinds of knowledge (to know how to know the knowledge, knowledge of how to be with a evidenciables actions, know what to do with the knowledge, know how to live as an ethical human being and share with others)
Metacognition
Differences
Goals, purposes and contents.
Knowledge and to know
Capacities, abilities and skills
Subdivision
Basic competencies
Generic competencies
Specific competencies
They are essential competences to live in society, such as writing, reading, arithmetic, and coexistence
They are transverse or common competences, to different areas
They are competences of one area, such as disciplinary or professional competences

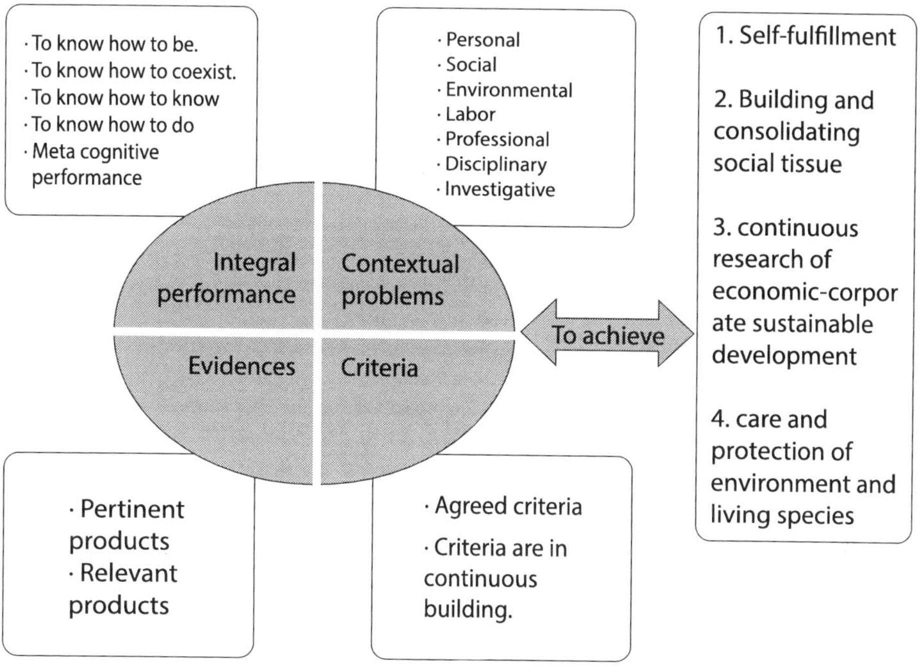

Figure 1.4 Central elements of socioformative conception of competencies.

Historic development of the concept of competencies

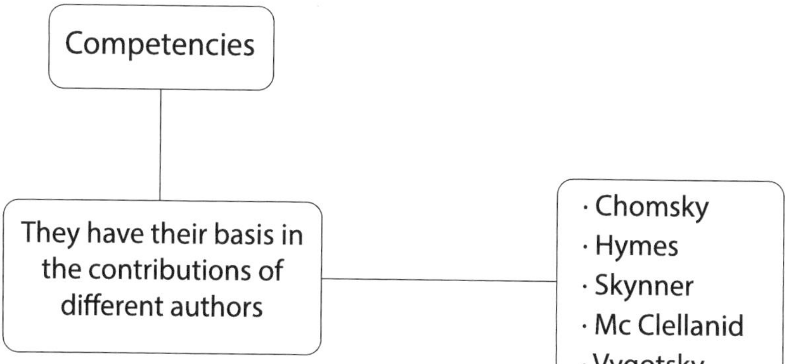

Which were the key sources in the current development of the concept of competencies?

The building of the current concept of competencies has its basis in the contributions of several authors as much as philosophy as in the education and in the organizational world. They tried to give answer to several problems related to the formation and performance, as it is described in the figure 1.5. This means that there is not an only single line under which the concept of competencies was developed (Echeverria, 2004.Tobón, 2005; 2006; 2012) but the lines are multiple and under diverse epistemologies.

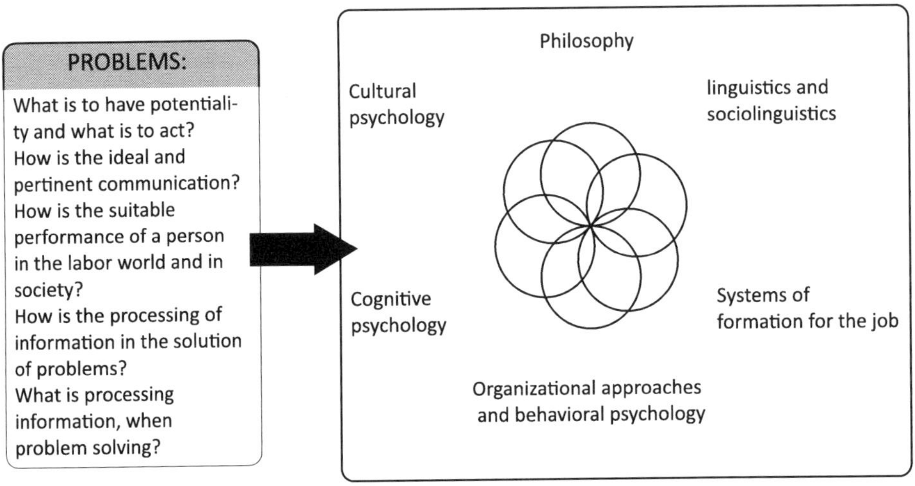

Figure 1.5 Main historical contributions to the development of the concept of competencies.

Next the contributions of some authors to the current concept of competencies will be summarized. They are framed on the figure 1.5.

Philosophy

Many current reflections about competencies have traditionally been set due to eastern and western customs; such as in philosophy, from the pre-socratic, going through socratics, until middle ages, modern times and contemporary era. These ideas have been set to understand the power and the act, the analysis of unity and diversity, the meaning of citizen, the analysis of talents, man's relationship with himself and the world, the reflection about to be or not to be, the search of knowledge, etc. Some of these aspects are done with great detail in some other bibliographic sources (Torres 2001, Tobón 2012a).

Nowadays, some of these philosophic traditional reflections occupy a central paper in theorization and application of the formation based on competencies. First of all, there is a question about what a good citizen is and his role in democracy. For this, there is a line of citizenship, competencies, a field of great development and work in Colombia.

Secondly, there is a line of unity in diversity and reflection about oneself and others which is the basis of working in generic competencies because these represent unity. From this, there are specific competencies whose essence is diversity.

Linguistics and sociolinguistics

Chomsky's linguistics. Noam Chomsky (1970) proposed the concept of linguistic competence as an implicit and genetically determined mental structure, which sets into action through the communicative performance (effective use of the linguistic capacity in specific situations).This author always opposed the framework of his generative transformational grammar competence performance.

According to this author, the competence is an internal cognitive heritage that the speaker- listener has of one language. In this sense the competence is common and identical in all the speakers of a single linguistic community. From this, the concept of competencies started to develop multiple critical elaborations, in linguistics, as well as in psychology (behavioral and cognitive) and in education.

Chomsky's concept of competencies still remains valid in authors and teachers who take on competencies as internal and implicit process and which, in order to be evaluated they require performance. This concept is opposed to what the socioformative approach proposes: that competencies have as implicit as explicit processes, where performance is part of the same competence and it is not any different thing.

A very important aspect to point out from Noam Chomsky is his conceptions of creativity in the competencies. He assures that competencies are multiple constructions that an ideal speaker - listener may do from some finite elements. This means that a competent person can address to different aspects of the problems with some deep and pertinently built, minimal knowledge.

Sociolinguistics. This line began to be developed by the late 1960's, and got consolidated in the 1970´s. Here the concept of competence is developed in the framework of people interacting with their environment and the communicative competence concept is proposed (Hymes 1996). It referred to the effective use of language and the linguistics in specific situations of communication, considering the demands of the context. In this form the competence is no longer an internal process and it begins to be integrated as performance in the context.

Hymes stated that knowing a language is a way of understanding, assimilating and assuming the world, and to use the language in the process of communication. In doing so, it is not enough to know the grammar rules, but it is also required to know when we must speak and when we must keep silent; where, how and which attitude to have. In this sense the communicative competence is diverse opposing to the linguistic competence which is homogeneous.

Nowadays, the communicative competence has a clear applicability nowadays, and it is an essential competence in basic and middle education as it is also in high school, in technical education and superior education. Now, there has been an advance in determining criteria with the goal to guide the formation and the evaluation of this competence, but its practice in all the educative levels is missing, because although there is progress in the understanding of communicative competence, there is still an emphasis on linguistic contents and very little effort to form competent people in language as it is the case with university students and professionals. To achieve this, it is necessary to work with project cases, role-plays, meaningful process of writing, essays, living techniques and strategies based on information and communication technology.

Systems of formation for work and labor education

Since the first decades of the XX century, the need to form people in the labor world began to be conceptualized, with the objective that the enterprises had qualified personnel in their different jobs. This happened first in the United States, Canada and different european countries, such as England. With this, a line of education for the job was developed. These facts became important in the 1970´s , when the interest for that type of education grew and when a major conceptualization of the need of having qualified personnel to develop some positions at work emerged. This was more than what had been taken into account by traditional education which was more focused on formation of academic knowledge. This is due to the growth of entrepreneurial competence and to the growing need of having qualified workers (Cariola and Quiroz, 1997).

This tendency continued in the 1980´s and in the 1990´s with great intensity and development; leading progressively to the constitution of the competence- approach as a key to have qualified personnel in the entrepreneurial organizations. This is how, from the 80´s on, in different countries of the world there have been establishing systems to strengthen the qualification of people in the industrial world, mainly based in four central elements: identification of competencies, standardization of competencies, formation of competencies and evaluation and certification of competencies in the world of jobs.

Three approaches of competencies have been considered (frame 1.5): structuralism, functionalist and holistic. Nowadays this tendency is practically led to use the holistic approach in the industrial world because each time there is a greater agreement that competencies must articulate, as much the structural aspects, as the functional aspects so that people become suitable.

This last approach is the one that begins to stand out in the international organizations which promote the improvement of qualification of workers. For example, the IBERPROF- OEI (2000) defines a competent person as that person who applies abilities, knowledge and skills in a wide variety of industrial contexts for this there must be agreed criteria of quality.

Frame 1.5 classic approaches of competencies according to the development vision.

Classic approach in the world of work	Emphasis
Structuralism	This emphasizes the components of competencies such as abilities, skills, attitudes, values and knowledge. -In this approach to set out a competence implies to do a detailed analysis of its components.
Functionalist	It emphasizes in the functionalism of competencies. It means in its contribu-tion to the requirements and challeng-es of the context and it's centered on the activities and tasks to do. Its core is to do as such and it gives little importance to the internal aspects of performing such as knowledge, attitudes and skills. Here the competencies are set out in a synthetic form, for example "the competence to research", or "the competence to set out" etc.
Holistic	It assumes the competencies as performances before activities and tasks from specific personal attributes such as abilities, knowledge and attitudes. In this perspective one must first see what the expected performance is, in the industrial context and then the diverse components of the competence are determined.

Organizational approaches and behavioral psychology

Studies about competencies in the organizations. In 1973, David Mc Clelland published several writings in which he launched the competencies to the organizational world, and to psychology. This author sets out that traditional academic tests of aptitude and knowledge as well as scholar grades do not predict the success in the work, and they have bias against minorities: women and people of low social-economic stratus . This led him to investigate about other variables and factors which may permit for a better determination of development of people concerning to labor. He found that to predict efficiency it was necessary to watch and to analyze to people in their job positions by comparing the quality of successful people with the characteristics of those who had an average development (Mc Clelland 1973).

This author built then the concept of competencies as a set of characteristics that lead to a successful performance. In this perspective, the competencies do not reflect all the characteristics of people in the industrial professional world, but only those that produce a successful efficiency (Mc Clelland, 1993). This led him to develop methods to determine variables of competencies that allow to predict the abilities at work. Some of the most important methods were: a) The sample of the use of criteria (comparison of successful people in the enterprises with less successful or of a mid-level people, to determine the related characteristics with success), and b) The identification of thoughts and behaviors towards the successful results.

Mc Clelland defines a competence as an underlying characteristic of a person, which relates by chance with the effective criteria- reference in the abilities, in a work or in a situation. These contributions gave way to diverse development and applications of the competencies in the organizational world, to which they have contributed other important researchers in this field, such as: Boyatzis(1982), Spencer and Spencer (1993) Woodruffe(1993) and Levy-Leboyer (1996).

Nowadays, Mc Clellands´ approach is a relevant theoretical referent in the management of human talent in the organizations, and his concept of competence as an underlying characteristic has transcended, because currently, the competencies are also assumed as explicit and observable processes in people, who give different advantages to the organizations.

Contributions to behavioral psychology. Opposite to Chomsky's proposal, who considered competence as something internal, and following the impulse to the concept of competencies given by Mc Clelland, in the line of behavioral psychology, the topic of competencies also began to be treated by the end of the 70's, with a better development in the 80's and 90's of the last century, by following the theoretical contributions of several authors. In this form the concept of competencies as effective, observable and verifiable behaviors of people began to be approached in the organizational framework. This proposal also integrates the development in the competence, such as Hymes did in the sociolinguistics, which differs from Chomsky's proposal. This way a solid behavioral model of competencies emerged, which is based on the relationship of people with themselves and with others, and the systemic functioning of organizations, as social as entrepreneurial. It is behaviorism which has transcended the stimulus answer scheme and which is based on the ecological and organizational functioning (behaviorism of third and fourth generations). The approach of competencies from the behaviorism has had notorious developments in the field of management of human talent in the organizations; in which, it is assumed the orientation to search that numbers of an organization to have key competencies for the companies to be competitive.

Since the mid 1990's, the behavioral conception of competencies has been implemented in educative institutions of several countries to form people with certain competencies which may give them a greater impact in industrial professional insertion, and to organizations, competitive advantages in the local national and international ambiance.

Re-Engineering. This area has contributed to deepen in the concept of competencies too, from the organizational corporate world by emphasizing in the generation of strategies in the organizations to have graded and qualified people who may answer to strategic goals of competitivety (Barbero, 2004). In this form, organizational management has in mind the work of competencies with the objective of having talent with required abilities, attitudes and skills for the company's processes to generate social and financial profitability.

Cognitive psychology

Another disciplinary line that has given meaningful proposals to competencies is cognitive psychology.

Theory of multiple intelligences of Gardner (1987-7997). This theory has shaken up the concept of intelligence and has given notorious contributions to understand competencies in its cognitive dimension (see Tobón, 2009a). Gardner (1995, p. 33) states that each...

> Intelligence implies the necessary ability to solve problems or to elaborate products that are important in a cultural context or in a determined community. The capacity to solve problems allows to talk about a situation in which an objective is followed, and to determine the path that leads to that objective. The creation of a cultural product is crucial in some functions such as acquiring and transmitting knowledge or the expression of the opinions or feelings. The problems to solve, go from creating the end of a story until anticipating the movement of checkmate, or to knit an eiderdown. The products

go from scientific theories until musical compositions, passing by successful political campaigns.

Straight on (p. 248) the author talks about competencies in the following terms:

> The capacity of people to acquire and to make knowledge to progress in a cultural specialty, as well as to apply it in a determined form directed to an objective. Key characteristics in some of the definitions of intelligence have to do with the resident competencies in the head of an individual as in values as in opportunities given by society to apply these competencies.

-This conception of competencies has evolved as in Gardner as in the framework of cognitive psychology in general. Today we understand that competencies and multiple intelligences are two different concepts but related between them. Multiple intelligences are general processes of processing information to adaptation and solution of problems in a determined cultural setting. Competencies by its side are processes of attainment of activities and/or problem solving with suitability and ethics which are based on one or several multiple intelligences. Thus, a person can have a high level of interpersonal intelligence, but a low competence in team, due to the fact that this person possibly lacks to develop abilities, skills and attitudes that are considered as essential in the criteria of evaluation of such a competence.

Theory of the cognitive structural change. The theory of Sternberg (1997) has been very important in understanding competencies, specially, its concept of practical intelligence, which refers to the capacity that people must have to develop themselves with intelligence in the life situations. This implies to get to a solution of the problems without a need of thinking and /or analyzing much with the objective of guiding the attention and memory in other activities. Colombia is one of the countries that has done substantial contributions to the concept of cognitive competencies, through diverse work and applications in the evaluation since the late 1990's of the last century (Torrado, 2000; Gallego, 2000).

Cultural Psychology.

In a different way, but also with a contextual approach, are the contributions of cultural psychology whose main representative is Vygotsky (1985) and which have been suggested by authors such as Torrado (1995,1998). In this line basically, it has been proposed that the competencies "are situated actions which are defined in relation to determined mediator instruments" (Hernandez et al, 1998. p. 14). They are situated actions in the sense that they consider the context in which they are carried out. Now, these actions come from the mind which is built in social relationship and it is updated by the culture (Vygotsky, 1985; Brunner, 1992).

In general, the cultural psychology has contributed to the concept of competencies the principle that the mind and the learning are a social construction, and they require of interaction with other people where suitability is influenced by the same context. (See, Tobón, 2006). Colombia has stood out because of its constructions around competencies from cultural psychology and the social constructivism that began at the beginning of the 1990's. In the educational practice there have been advances, but there is still a lack of many strategies to implement from the perspective of cultural psychology.

How do different historical sources converge in the current concept of competencies?

Many authors criticize the concept of competencies because of their multiple definitions, approaches and emphasis as well as of their great amount of sources in which their development is based (see, for example, Bustamante, 2003). However from complex thinking, this is more a fortress than a weakness of the concept, because they give different lines of deepening that make it flexible and connect it to different academic and social traditions.

This is one of the contributions of complexity in the building of scientific theories, unlike traditional positivism, which only looked forward unambiguous concepts. This has one single meaning. This does not mean that all definitions of competencies are valid; on the contrary, pertinent definitions must have the same elements of base, although its wording and emphasis vary.

The building of a compressive and integral concept of competencies must consider the different historical sources including the intellectual ones in a trans disciplinary framework. This is the line that has been followed in CIFE Institute since 1998, based on complex thinking and articulating the contributions of several authors and disciplines. This has allowed to reach a trans disciplinary concept of competencies (Tobón, 2012a), as a key axis of socioformation.

In the several experiences of application of the approach of competencies in industrial organizations, colleges and universities of several countries of Europe and Latin America, it has been possible to determine what is precisely this trans disciplinary character of competencies. This allows greater possibilities of communication and articulation among the diverse disciplines, social organizations, enterprises, epistemologies and human talent management policies.

This is why one of the main tasks of the socioformative approach at the moment of implementing the competencies in a determined organization, is to articulate several referents, such as old and new ones which help to a major impact on the established goals. And this is also an invitation and recommendation to all the enterprises and educative institutions that are working on this topic; so that they conceive and apply the approach of the competencies, by articulating in a systematic form and in webs several contributions in the area, from pertinent and coherent models with the institutional philosophy, based on a determined vision of the human being and society.

In this sense, in the complex line, then, the trans disciplinary work is essential to create integrative, coherent and pertinent models; concerning to competencies and its application in education and in the world of the organizations.

Categorization of the competencies

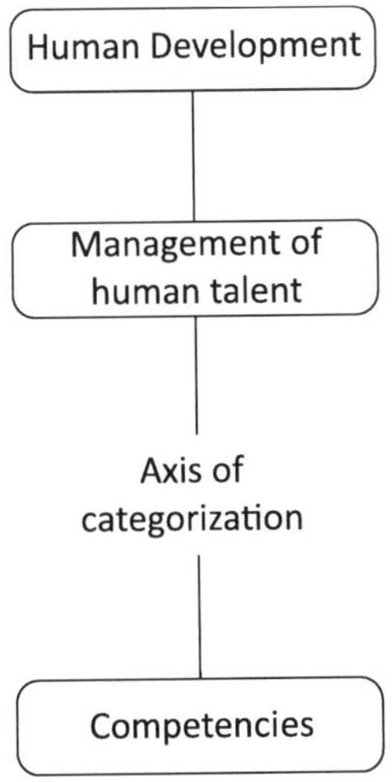

The competencies are classified in the management of human talent and this, on its side, in the human development in general.

This last idea is known as the process by which humanity must work on a human scale (Max-Neef, 1993, 1996) to reach levels, each time with higher life quality in all the areas, in the framework of search of meaningful progressive advances in the social justice and coexistence and to create sustainable relations with the ecological environment with concrete actions to diminish current, serious environmental problems and to prevent future problems.

In relation to the management of human talent, we must indicate that this concept is not equivalent to the "handling (or management) of human resources" because when "human talent" is set out the emphasis goes directly to detect continuous zones of talent in people and to develop at the same time those zones of talent to have competent people in the organizations including the educational ones and in this form to look for sustainable human development. On the opposite side, the concept of "human resources" does not have in its origin the character of formation neither does it have the character of continuous perfection, it rather emphasizes in having people to answer back to requirement of job as we have other resources such as capital, the technology, furniture and essentials, etc. In this form the " human resources" are equated with other resources of an organization and this is a reductionist static and fragmented vision of human beigns, especially if it is considered that they are the key of any organization.

What is human talent? The most accepted definition is that human talent refers to the capacities and attitudes set into action with commitment. A person has talent when he sets into action with commitment and sets several conditions to obtain determined results, valued as relevant ones in the context.

For example, when an architect designs a new type of a house with original characteristics which are out of the conventional parameters we say this person has talent. Equally, when a journalist investigates a human problem with perseverance when there was little knowledge of that problem and he achieves to communicate it clearly we say that journalist has talent for investigative journalism, and the person who plays tennis skillfully and obtains many victories in tournaments and competitions we say he has talent to play tennis and so on. Then the talent is diverse since it sometimes implies to investigate novel aspects, and other times to compete with others.

Human talent management does not circumscribe solely to the social or entrepreneurial organizations: This is a concept that is also used in education and in the social formation to refer to the global process by which some actions are implemented so that people develop their talent and they put it in practice in several contexts according to determined goals.

The competencies express in a concrete form the human talent developed and supported effectively. All this in a framework of determined contexts: social, labor, professional, disciplinary, personal and investigative. Likewise, the competencies make it possible to manage the human talent clearly as in the organizations as in educative institutions when they offer conceptual and methodological tools to look to form, to evaluate to certify and to promote the expected human talent (Fig 1.6).

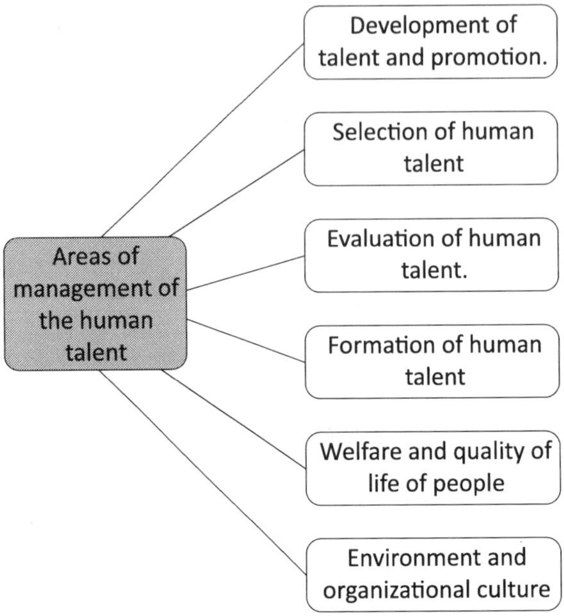

Figure 1.6 Human Talent Management by Competencies.

The importance of establishing programs of management of human talent rigorously, in the organizations is shown on the Pfeffer´s study (1999). This researcher reviewed over 130 studies, between 1961 and 1991, about organizational changes, going from the strategies of management of human talent and, he found an economical, meaningful increase in the companies which confirms the importance in this field in nearly 75% of the cases .

CHARACTERIZATION OF Competencies

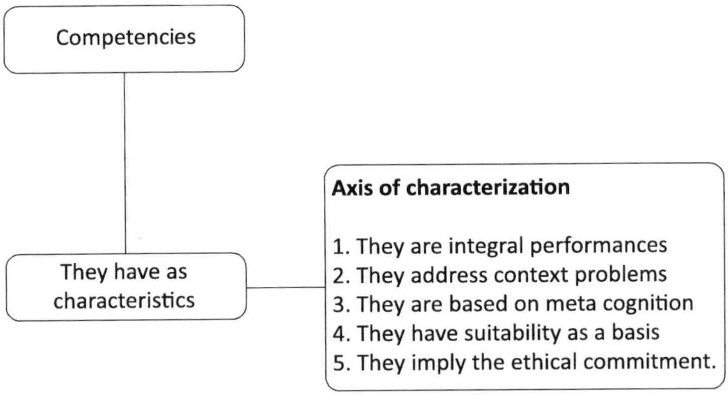

What characterizes a concept is all that gives it an identity as an academic scientific element. In the case of competencies, the advance has been very little to agree widely its essential characteristics in order, to give identity and coherence to this concept, as well as to facilitate its understanding and use in the management of human talent.

The term competence is analyzed from its historical development and main use of the concept today. It can be said that is has the following main characteristics:

1. They are integral situations.
2. They talk about problems of the context.
3. They are based on metacognition (continuous improvement).
4. They have suitability as a base.
5. They imply an ethical commitment.

In the socioformative approach, the processes occupy a prevailing place in all the curriculum design,in the didactics and in the evaluation. What are the processes and what do they differentiate from? The processes are articulated actions that emerge from entrance information to reach determined results in an organizational and/or ecological environment through several resources. They differentiate from the work because of the functions, activities and tasks; aspects which are more concrete and less dynamic.

The competencies, when conceived and tackled as processes are dynamic and contextualized. They constitute actions of processing information, knowledge to reach determined goals.

This means that curriculum design is based on the analysis of processes, more than in the study of punctual tasks; which lose applicability rapidly. However the processes of performance are complemented with the emphasis on concrete products that limit them in a specific space of time.

Next, the key characteristics of competencies are described based on socioformation.

Integral performances

An **integral performance** is a performance in a determined context with sense, which applies in an articulated form several knowledge not only for the person who executes the action but also for the surroundings with pertinence and power of verification. For example, the competence of leadership for the person who develops it implies to apply actions to orient people and/or teams of work.

The knowledge (to know how to be, to know how to coexist, to know how to do and to know how to know) is applied in an articulated and coherent form with the objectives of performance according to the requirements and challenges of context and with the own planning of the person (frame 1.6). In the process of curriculum design, to know how to coexist has been integrated in the know how to be, to facilitate the planning of the learning process. The thesis of the socioformative approach is that every competence is set on action based in the integration of these knowledge, from the framework of metacognition.

Frame 1.6 Summary of knowledge of competencies.

To know how to be	It refers to perform with attitude and values, with a sense of challenge and motivation. It considers the ethical project of life in an intention of continuous improvement.
To know how to coexist	It means to be in contact with other people with cooperation and assertive communication, in the framework of respect and responsibility.
To know how to do	It consist on setting procedural abilities and pertinent techniques into action to a determined proposal and context, and to know who improves these and to argue their qualities too.
To know how to know	It is based on having abilities of thinking concepts and categories to process the information of context and to generate knowledge, in articulation with other knowledge, in a pertinent form.

Resolution of problems of different contexts

The competencies imply people to identify, to interpret, to argue and to resolve problems of different contexts. In this proposal there are 2 concepts which need to be clarified: problem and context.

The problems are challenges of how to pass from a given situation to an ideal or a desirable situation then. They imply to identify a need and to satisfy it.

The contexts are surroundings in which human beings live and which give a meaning to life. The most relevant contexts are: the personal area, the family, the social institutions (for example, the religious confessions, the army, etc.) corporate organizations, cultural expressions, laboral - professional world, the science and the ecological environment. The competencies are set on action in one or in several contexts with sense and pertinence. They look for an achievement of determined purposes by assuming the challenges of present and future.

Metacognitive process

The Metacognitive process is essential in the socioformative conception of competencies. It is conceived as the continuous improvement in the performance to achieve determined goals, based on reflexion. Therefore, it is not enough to take consciousness of mistakes: it is essential to overcome them to achieve the pertinent and valuable goals (Tobón, 2012a). In the classical theory of metacognition (Flavell, 1987) this is understood as to take knowledge of several factors that influence performance (related with the person, the activity, the strategies and the surroundings) and then to self-regulate them. It means to plan to monitor and to evaluate to ensure the fulfillment of the goals in the performance (fig. 1.7).

Figure 1.7 Model of metacognitive processing in the competencies.

Understood in this form, metacognition is a practice that has to be worked in every competence, and it must be present as in the mediators of human talent management (for example teachers), as in people in the process of formation, evaluation, certification and promotion.

Suitability

Suitability is the grade in which performance fulfills with determined quality criteria. It is to do the activities or to resolve the problems by fulfilling with established criteria of efficacy, efficiency, effectiveness, pertinence and socialization, established for this effect, with argumentation and socialization (frame 17). This is one essential characteristic in competencies, and it marks their difference with other concepts such as the capacities (in their structure suitability is not present). This means that a competence exists if the person shows a suitable performance. It means with pertinent criteria.

Frame 1.7 Differentiation of essential concepts in suitability.

Efficiency	It is to reach the maximum results with the minimal resources, human talent and time.
Efficacy	This is the achievement of expected results in the process of performance. It is evaluated by comparing the quality of the obtained results with what is expected. It is necessary to consider errors and unwished results.
Effectiveness	It refers to the process of quantifying or determining the achievement of established goals in the process of performance.
Pertinence	It has to do with the performance and the results which have to correspond with the expected one in the context and /or in the own goals of the person.
Relevance	It means that performance in the competencies must be valued as representative and important in the context so it can be suitable.

Ethics

It consists on acting with universal human values in the different situations of life, such as responsibility, respect, justice, life, truth and honorability. This is essential to achieve personal self – fulfillment, social coexistence, quality life and social sustainable economic development as well as environmental – ecological equilibrium.

In the competencies, ethics implies specifically to analyze before acting, the consequences of the own acts. It is to have them according to universal values, to answer back because of the consequences once one has acted and to correct the errors as soon as possible, which implies to repair possible damages on other people or on one self. The principle in the competencies is then that there cannot be suitability without personal and social responsibility because to form in competencies is to form citizens for the life world. (In the academical, laboral, professional and everyday life) who must assume a critical attitude before each situation (Bogoya, 2000 p. 12).

Differentiation of competencies

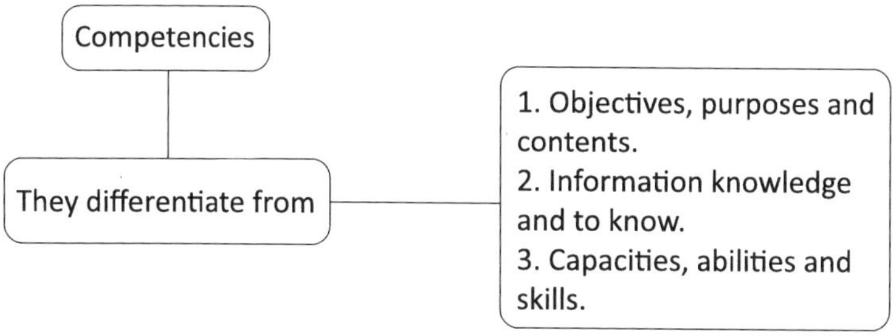

How are competencies differentiated from objectives, purposes and contents?

One of the major obstacles to implement the management of human talent by competencies in education and in the organizations is the lack of clarity that still exists towards objectives and the purposes; two forms of educative planning and evaluation of learning which have been the paradigm of formation for more than three decades.

Next some guide lines to contribute to this difference are set out and in this form to achieve the implementation of competencies.

The approach of the objectives is oriented to determine the concrete behaviors that the students must achieve when they finish a well delimited period of study. It is set out by considering the academic contents and they are proposed for each knowledge based in some determined taxonomies as for example Bloom´s Taxonomy.

The purposes are also goals of learning but they differentiate from objectives because their emphasis is set mere on the didactic process and in the evaluation. Generally they are established from the teacher's perspective. They follow a more constructivist approach than the objectives which tend to be more conducts.

Different from the objectives and the purposes, the competencies are not centered in concrete behaviors, which are intended to be achieved by the end of a period of study. They are not centered in the goals of learning considered from the process of formation either. Rather they are integral performances to solve problems with suitability, ethics and continuous improvement which articulate different types of knowledge and they must begin by being shown since the beginning of the formation process. Therefore, the evaluation does not consist on determining if the students have acquired determined behaviors. Rather it determines how they perform before the challenges of the context and how they demonstrate this through relevant products (frame 1-8).

What are information knowledge and to know about competencies?

Ortega and Gasset (2004) has set we wish "we knew but however for millenniums the man has worked to know and he has achieved very little knowledge. This is our privilege and this is our dramatic condition" (p. 107). Traditionally, human beings have looked for producing immense amounts of information but uncaring about production of knowledge as well as conversion of knowledge in to know. This is due to the human tendency to confuse knowledge and to know with information which generates conformity because the information does not imply critical analysis, transversally contextualization and responsibility.

Frame 1.8 Differences among competencies, objectives, purposes and contents.

Components	*Example*	*Competencies*	*Example of competencies*
Objectives: Expected, concrete behaviors by the end of a period (for example at the end of a bimester)	**Objective 1.** To identify the problem in a research project. **Objective 2.** To analyze the most pertinent method to tackle a research problem. **Objective 3.** To evaluate the results of the research process.	The competencies are not concrete behaviors, but integral performances. They do not have to appear at the end rather, the students must begin to show them since the beginning a determined level of control.	To resolve problems of the context through a determined research methodology to generate knowledge and to perform with impact on reality by considering the accumulated knowledge, collaborative work and ethical commitment and writes.
Purposes: They are goals that the teacher proposes to himself in the didactic activity, by considering the purpose of learning and constructivism.	**Purpose 1.** The student understands what assertive communication is, through socio-drama so that they apply in the everyday life. **Purpose 2.** The student learn how to plan the purpose of reading and applies in everyday life through rehearsals	Competencies are not the coals of the teacher but concrete performance of students before problems. They are integral performance before problems of social, labor-professional and disciplinary contexts, with suitability and continuous improvement.	To use the oral and written language to communicate with understanding in varied, social, cultural contexts, by using different codes and tools in the frame of a metacognitive process.

Components	Example	Competencies	Example of competencies
Contents: They are topics to be addressed in the subjects and different formative spaces, given from the systematization of knowledge in a determined area of knowledge (see Tobón, 2012d).	Carrier: Chemical Engineering Subject: Mathematics I academic period: semester I. Subject 1. Systems of lineal equations - Lineal systems - Gauss method - Theorem of Rouché-Fro benius - General method for the resolution of a system of lineal equations. Topic 2. Matrices determi-nants. - Matrices - Operations with matrices. - Inverse matrices: Gaus-Jordan method. - Determinants. - Properties of the determi-nants. - A formula for the inverse matrix. - Special determinate.	In the competencies the contents are a resource and a means for the formation and boarding of challenging situations. But they are not the purpose of education as it happens in traditional pedagogy.	Competence of mathematics: Resolve problems thorough mathematics with the objective of reaching determined goals in the context by applying the metacognitive process. This competence is an integral competence before contextual problems that implies the management of mathematical contents as well as other resources.

The information gives data of a determined reality which are processed by the human mind and they are shared in social spaces. The knowledge, on the other hand, is a set of interwoven representations which is based on the processing of information, with analysis, synthesis, interpretation and reasoning, in a determined context, with meaning and consciousness of its interrelations. On its side, knowledge consists on having a wide set of knowledge to apply them in activities and problems, considering its implications from an ethical commitment (Tobón, 2012a). Knowledge is linked to wisdom, it implies to use the knowledge with ethics on favor of human, social and environmental development in a sustainable and equilibrated form.

The competencies are processes that integrate knowledge from a suitable framework to do activities and to resolve problems with a sense of challenge and continuous improvement. In this sense, knowledge is based on information; the knowledge into wisdom, and competencies in knowledge. (fig. 1.8).

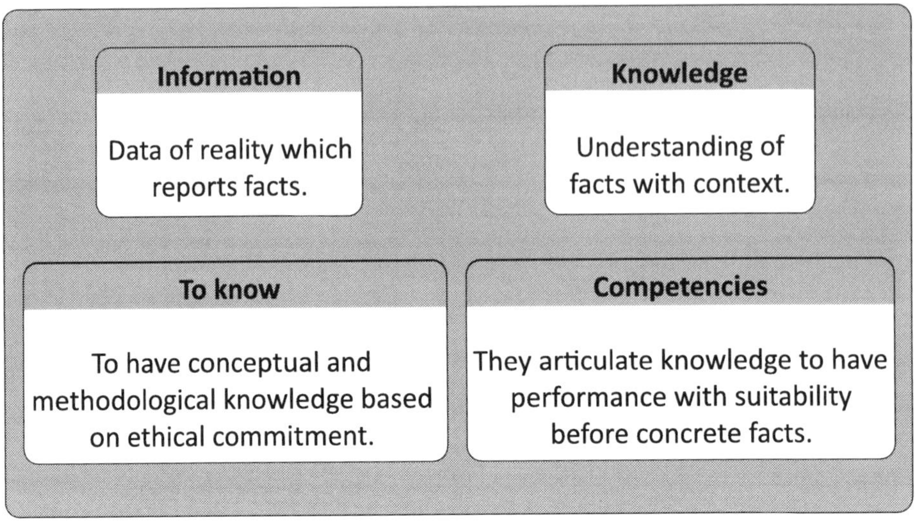

Figure 1.8 Differences among competencies, information, knowledge and to know.

The knowledge and to know, as they have been presented in some lines before, have then a complex nature; they are multidimensional, contextual and evolutive processes. That is why to manage the knowledge from this understanding it is required that people develop a complex way of thinking that allows them to do so because knowledge and to know constitute our way of thinking.

In traditional education it has been insisted on the formation of a simple thinking, and that is the thinking that has been supreme in society which has led to equate information with knowledge until wisdom. The current challenge is to advance towards wisdom from multidimensional knowledge. We must face the change of thinking in our mind by passing from the simple way of thinking, and from the complex thinking to the easy one.

How are competencies differentiated from capacities, abilities and skills?

Although there has been a very great advance with respect to the understanding of competencies in education and in the organizational world, there is still a little confusion of competencies and capacities, abilities and skills. It happens as in countries with advance in the implementation of this approach (Finland, Australia, England, Colombia, Chile, and Mexico among others) as in international, current projects now applied (for example, the Alfa tuning project, in which competencies are frequently described as capacities and skills).

In the frame 1.9 a synthesis of the main differences among competencies, capacities, abilities and skills is shown. Basically the major difference is that competencies are integrative performances (composed of knowledge, metacognitive processes, values, attitudes, capacities, abilities and skills) are not interrogative, nor do they tackle metacognition or suitability. Although it is important to understand these aspects in every competence, it is not precise to explain them in the curriculum or in a formative project.

Frame 1.9 Differences among competencies, capacities, abilities and skills.

Concept	*Definition*	*Example*
Competencies	They are integral performances with suitability, metacognition and ethics that articulate knowledge from the metacognitive process.	Competence of leadership.
Capacities	They are general and internal potentialities. They do not imply metacognition or suitability. The competences are developed, based on capacities.	Capacity of establishing a relationship with other people.
Abilities	They are actions to do some activities with the purpose of reaching determined objectives. They are developed from capacities.	Ability to reach goals with people.
Skills	They are concrete tasks to set each ability into action	1. Skill to identify a goal. 2. Skill to write down a goal. 3. Skill to agree goals with people.

CLASSIFICATION OF Competencies.

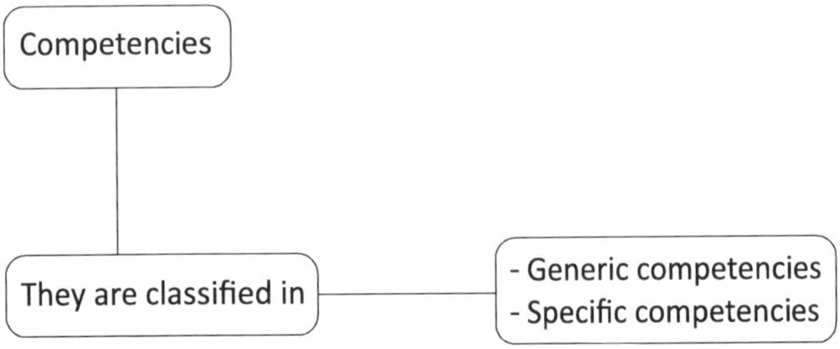

From the socioformative approach, three kinds of competencies are proposed as basic, generic and specific competencies. This is essential to articulate the educative cycles and to facilitate the recognition of titles as well as the mobility of students, professors, researchers and professionals. However it is known that the process of reform of several Latin American and European countries sets out other types of additional competencies as it is the Colombia´s case. It is a country that addresses, besides basic, generic and specific competencies, the citizenship and labor competencies. This also happens in Mexico, where the Reform of Education in all the levels is oriented by competencies and in which the disciplinarian competencies are considered in a complementary form, and the professional competencies in middle superior education. These adaptations by some countries show different contexts, needs and challenges.

Concerning to diverse kinds of competencies the essential thing is that we are before differentiated scopes of development. The generic competencies are based on the performance in different context; specific competencies are a performance in concrete and specialized environments and basic competencies on their side, are the base as for generic competencies as for specific competencies.

What are basic competencies?

They are defined as minimal or essential performances that all the members of a society must possess to perform in their life with suitability. In this sense it refers to strictly necessary competencies to live in a community. In turn they are the base for the development of other types of competencies, such as generic and specific competencies because they also have very concrete elements and they also have transverse components.

Generally basic competencies are developed in basic education (which includes preschool, elementary school and junior, high school), that is why it is mandatory, as it is included in most countries of Latin America. However in recent years it has also been stated that middle- superior education must be mandatory. In figure 1.9 the relation among several kinds of competencies in the educative cycles is shown.

They are examples of basic competencies:

- Reading competence
- Writing competence
- Arithmetic competence
- To coexist with others competence

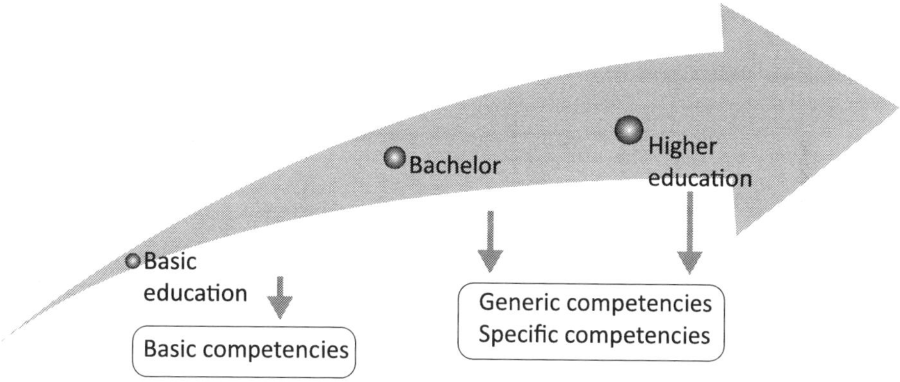

Figure 1.9 Competencies in several educative cycles.

What are generic competencies?

They are transverse competencies, common to different people, areas, contexts, disciplines and professions. They are not related to a specific disciplinary field, although they are based on the knowledge of diverse Academicals areas, as well as on life experiences of people and the challenges of the context, in a global framework, from the knowledge society.

It can be declared that generic competencies are more important than specific competencies related to a concrete and differentiated area because they let to face the processes of change in the world and they provide the faculty to people for management, creativity, entrepreneurship and innovation, based on collaborative work, ethics, clear and assertive communication and the management of technologies of information and communication. And this is what entrepreneurship (governments, social leaders and researchers) is demanding with big emphasis.

In a more concrete manner the reasons to tackle generic competencies in the plans of study are the following ones:

1. They increase considerably the possibilities of mobility between different positions of job because people can make more pertinent contributions to the organizations and they can be in a process of continuous improvement.
2. They facilitate the acquisition, development or consolidation of specific competencies to perform in new processes in the life.
3. They promote mobility in a country and in different countries, because they facilitate a comparable and compatible performance, which is different from homogeneity (see Gonzalez and Wagenaar, 2003) because they do not look for totally equal competencies, but competencies with common situations.
4. They make possible the acquisition development and/or the consolidation of specific competencies of other professions to increase the possibilities of professional insertion.
5. They strengthen the possibilities of employability and labor correlation of professionals, because they are competencies which are more requested each time by organizations, due to the fact that they generate major adaptation to changes and facilitate innovation, continuous learning, work by nets and collaboration, among other aspects.

In diverse projects, generic competencies of great relevance for education have been established, which are considered as referent in a framework of processes of construction and management of curriculum, and also in didactics evaluation and certification of competencies. Some of these projects are described on frame 1.10.

What are specific competencies?

They are competencies that give identity to each one of the disciplines and professions, by forming a basic core. For example the competence of an architectural design is typical of the architecture, the competence of psychotherapy is typical of clinical psychology, and the competence of auditory of systems is typical of engineering of systems. The specific competencies are determined with base on different methodologies, such as analysis of problems in the surroundings and the study of disciplinary, social and organizational process, in each one of the professions or occupational areas, having in mind the dynamics of present and the tendencies towards the future.

Frame 1.10 Some international projects with generic competencies.

Project / Element	Eurodyce project	SeDeCo Project	Alfa-Tuning Project	European Framework (European Union,2009)	CIFE Institute Project (Tobón, 2012 a)
Purpose	To establish the vital competencies for a successful participation in the society which must tackle mandatory education.	Identification of key competencies for a prosperous life and a society with good functioning.	To agree the generic, necessary competences for quality education and internation al mobility in Latin America.	Key competencies that everybody must be able of acquiring from which depend any type of successful result.	To identify the most important, generic competencies to be working on in all the educational levels, for the management of human talent and entrepreneurship.
Purpose	To establish the vital competencies for a successful participation in the society which must tackle mandatory education.	Identification of key competencies for a prosperous life and a society with good functioning.	To agree the generic, necessary competencies for quality education and internation al mobility in Latin America.	Key competencies that everybody must be able of acquiring from which depend any type of successful result.	To identify the most important, generic competencies to be working on in all the educational levels, for the management of human talent and entrepreneurship.
Methodology of investigation.	Interviews, analysis of information and meetings among academic people.	Definition and selection of competencies of OCDE, with base on field studies, interviews and academic agreements.	Interviews with students, graduated employers and professors.	Bibliographic studies, analysis of European tendencies, and agreements among researchers.	Documental research, field studies and interviews with experts.

Project / Element	Eurodyce project	SeDeCo Project	Alfa-Tuning Project	European Framework (European Union,2009)	CIFE Institute Project (Tobón, 2012 a)
Number of proposed, generic competencies.		Key competencies are grouped in three categories: 1.- to use the tools interactively. 2.- to interact socially in groups. 3.- to act with autonomy.	27 generic competencies grouped in three categories.	8 competencies: - Mother language communication - Mathematic competence and basic competencies about science and technology. - Digital competence. - To learn how to learn. - Interpersonal, intercultural, social and civic competencies. - Entrepreneurship competence. - Cultural expression.	10 competencies: - Self-management of formation. - Oral and written communication. - Oral and written communication in a second language. - Work team and leadership. - Management of information and of knowledge. - Resolution of problems with base on mathematics. - Resolution of problems with base on natural sciences. - Entrepreneurship. - Research - Management of quality.

Connection of the COMPETENCIES

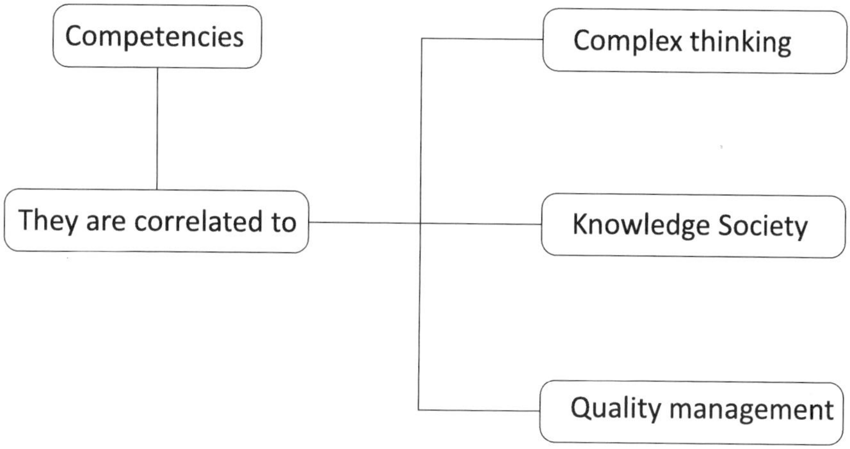

What is complex thinking?

In the 1980's, new epistemological approaches to understand science, management and knowledge began to appear. These approaches, such as autopoiesis, the adaptive complex systems, the chaos theory, dissipative structures, and fractal geometry, among others, are framed little by little in a general field of complexity science. All that means a change of paradigm in the sense of Kuhn (1962) for the understanding of science, although, the traditional science based in laws certitudes and lineal relations of cause – effect this approach is still dominant.

Complex etymology is the convergence of the proposals of different researches, among them several Nobel Prize, such as Murray Gell – Man (1994) and Ilya Prigogine (Prigogine and stingers, 1984, 1997). It is also important to outstand the work of Arthur (1995), Holland (1996, 1998) and Kauffman (1993, 1995). In addition, there are Morin's contributions (1995, 2000 a, 2000 b, 2002 a) who articulates complex thinking with the understanding of the nature of knowledge, Trans disciplinary, antro – poetics, and the Tierra Patria. Finally, it is important to consider the contributions of Maturana and Varela (1994) about the live systems and autopoiesis.

All the phenomena are studied in the complexity, over all, those which present non lineal dynamics and emergent properties, named complex adaptative systems (SCA). According to Begun, Dooley and Zimmerman (2003) a SCA is a system in coevolution which seeks self-organization. It is structured by diverse parts and it considers the context and learns from it.

Complex thinking as such, is on the line of theories of complexity, and was proposed by the French Professor, Edgar Morin. Complex comes from complexus, Latin word that means "tissues among tissues" (Dominguez 2006). Therefore complex thinking as epistemology is a method for building knowledge of any phenomenon, based on the form and dynamics in which such phenomenon is woven and with respect to other phenomena, with the objective of understanding it and to explain it in its process of order – disorder – reorganization, through disciplinary, multidisciplinary, interdisciplinary and trans disciplinary analysis. From this in the complex thinking the knowledge and to know are always multidimensional, transversal, changing and evolutive.

In the building of knowledge of reality, it is precise to take into account that this is not simple or complex: Reality is what our thought is. If thought is simple, reality is going to be simple. If our thought is complex, then the reality is going to be complex (Morin, 2002 a). This involves knowledge directly, because, with a simple thought, which primes in the society, it won't be possible to build pertinent and meaningful knowledge, which helps human welfare in a sustainable development in ecological equilibrium.

The performance in the reality from the competencies is more than a matter of capacities and cognitive abilities. Its essence goes through simple or complex thought that a person has developed. That is why we see, for example, people without academic studies or with cognitive disability who build knowledge in a meaningful and pertinent form; because they possess a complex thinking, whereas there are professionals whose building and management of knowledge are rigid and little meaningful because of the marked presence of a simple thought, given by rigid schemes of thought, in spite of the fact that they have major cognitive capacities than people with cognitive disability.

According to our criterion this is the major inheritance of professor Morin: the teaching that the problem with the knowledge is not in the reality not even it is a construction of itself; neither is it in the methods of knowledge, but in the self-thinking that orients the observation and knowledge of reality. Our understanding and its grade of depth, as well as its ecological sense, depend on the way in which we orient about knowledge. It also depends on the presence or absence of rigid schemes and on our effectiveness towards reality. This is not a matter of capacity, nor is it of ability, but about flexibility and commitment when facing the world.

What is to think with complexity? To answer this question, it is necessary to understand that to think in a simple form is to assume things in a lineal and one-dimensional form. It is to give little space to change and to uncertainty. And this is what has taken precedence traditionally in the management of information processes and of those of knowledge. On the other hand, to think in a complex form, according to Morin´s answer (1995, 2000a, 2000b, 2002a, 2002b) and of other representative authors in this field (see Tobón and Nuñez 2006), it is to watch to understand and to build reality with multidimensionality, flexibility, pertinence and evolutive sense by appreciating the continuous interrelation of processes of order with the processes of disorder and change (fig 1.10). Thus, it is possible to get the boarding of things in a more simple form which is different to simple.

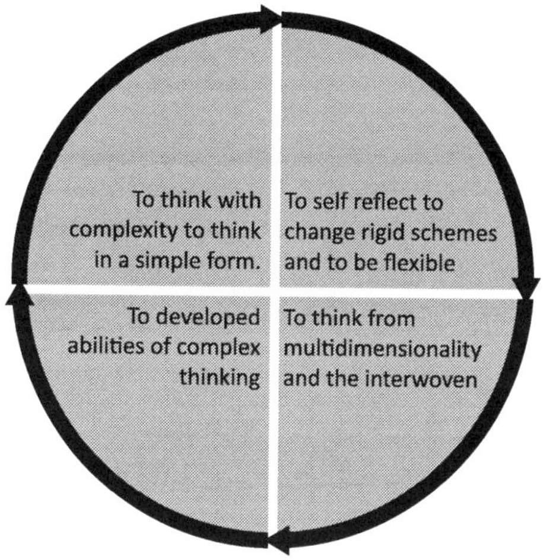

Figure 1.10 to think with complexity.

This is the challenge in nowadays knowledge society: To get the development of complex thinking in people, by changing and transforming the traditional educational structures that have prioritized formation, most of the times in a form of simple thinking. This is essential to advance in the building of meaningful knowledge from the enormous amounts of information that we have in society day by day and that many times we do not use to build welfare and life quality. The challenge is to get to a true knowledge that allows people to develop as human beings to participate in the building of social tissue and to look for equilibrium with the environment (figure 1.11).

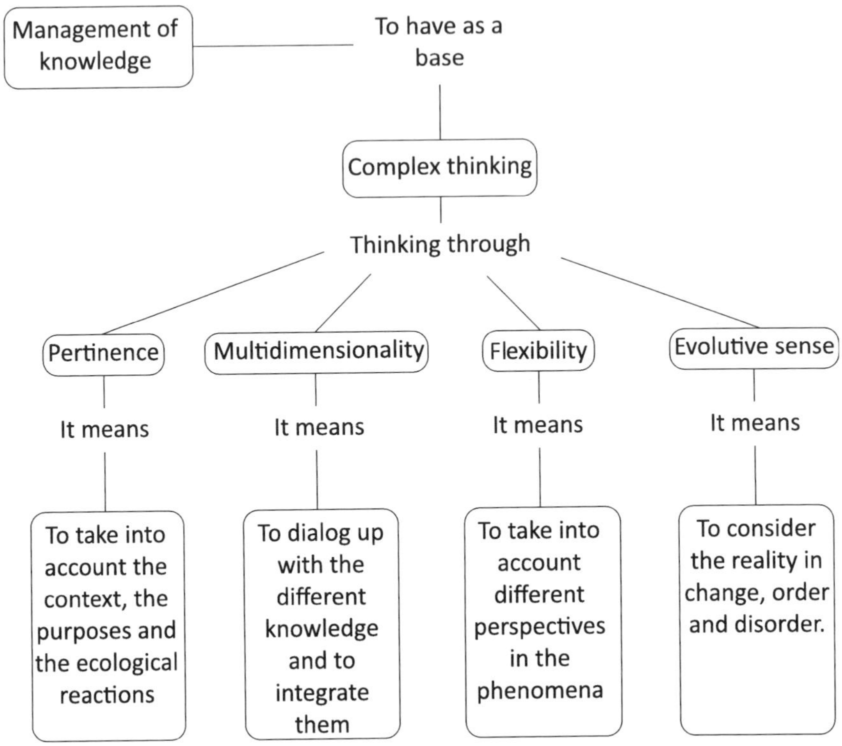

Figure 1.11 Central aspects of complex thinking.

Which are the abilities of complex thinking?

The complex is an interwoven of knowledge in the frame of multidimensionality and evolution (order- disorder- reorganization).

The competencies are complex processes because they imply articulation and application by knitting diverse knowledge and human dimensions by considering the situations of context and the challenges of the moment because its set in action many times involves facing uncertainty.

Additionally, complex means to think with complexity and for this the development of abilities of complex thinking is required such as metanoia, flexibility, dialogic etc. To deeply understand the competencies and then to apply them with suitability it is very important to develop abilities of complex thinking due to the fact that these abilities facilitate the understanding of context they guarantee creativity and innovation and they are essential to act ethically, according to a what for.

The development of competencies for human and social development requires people to get formed in abilities of complex thinking, whose set is what professor Morin refers to as complex mind. This is one of the major contributions of complex thinking to education nowadays, and it is the essence of the complex approach of competencies. Here is each one of these abilities summarized:

Evolutive ability

This is the ability to think about the phenomena in their processes of change, by establishing how there is a continuous organization- de-organization and re – organization in them, according to the systematic relations with the context. This implies to look for factors of order and organization in each phenomenon, also considering possible problems and uncertainties, and to establish strategies to face them in the present and in the future in favor of the human- ecological sustainable development. To develop this ability that we, human beings, little guarantee and use, it is required to have the sensibility with the own evolutive process to learn from this and to apply such learning in other processes of life.

Metanoic ability

It consists on the process by means of searching at least three approaches, methodologies or different perspectives to tackle a subject, with the objective of generating a new understanding and to create and to innovate a strategy to take it , and to consider the context in its systematic organization. This requires commitment, initiative and a strong motivation of achievement, because one must be persistent in innovation, and at the same time to have much strength to overcome rooted perspectives that sometimes block the process of creation. When there is only one model, a new development or change is sometimes difficult. That is why metanoic ability is essential to generate new proposals of human and environmental wellbeing in a systematic equilibrium whose essence is the change of mental models that make the dynamics of processes difficult. According to this, some pertinent strategies of change of mental models can be found in Senge (1995).

Metacognitive ability

This is the ability to become aware of our thoughts, knowledge, affections and performances, as well as exterior reality, to self- regulate such processes in favor of a better performance, understanding, ethical commitment, human development and environmental equilibrium. In summary: Metacognition is self- reflexion about our own performance and then to self- regulate for a continuous improvement. This is the best ability in complex thinking to invigorate personal change and in reality. Additionally, it constitutes the essence of competencies in the frame of complex approach as it is shown in the figure 1.3 of this chapter.

Hologramatic ability

This is the ability to understand reality by means of the analysis of how everything is organized in each one of its parts, of a determined process (personal, social, ecological, economical, technological and/or entrepreneurial)and this does not solely apply to analysis and understanding of reality; this is also a principle to go on in every performance looking, forward that everything remains configured in each one of its parts of the processes. Therefore, in this way, a full systematic pertinent and coherent articulation exists in the frame of achievement of determined purposes. In this activity, to consider the great goals of humanity results quiet relevant, so that everything that we do goes in this direction, and therefore, we can aspire to resolve critical problems that affect us, such as pollution, the destruction of natural resources, lack of social justice, etc.

Dialogic ability

This ability consists on developing, at least two actions. The first is to determine in the processes opposed, dissimilar, contrary or contradictory aspects, and then to look for complementing them in one, several or all its characteristics, in such a way that this generates new dynamics, that contribute to generate new possibilities of being and of doing, to overcome conflicts and difficulties, and with this, to contribute to a better quality of social and environmental life. The second action is implemented when in the processes there are not clear counter parted aspects. It consists on implementing in the understanding or in the processes, opposed factors that complement to dynamize reality and to generate processes of creative and innovated changes. In summary: the dialogic ability is to generate creative processes through the search of complementarity of opposed factors.

Ability of flexibility

It is the ability to assume, to face to search or to administer the change in one self, in others and in the environment with open mentality, agility and appropriate handling of the resources and of time, in favor of goals of human social and environmental development in systemic relation. A person has flexible thinking when he assumes the change and faces it strategically with the minimum resistance and a positive thinking.

Self- organizing ability

It consists on looking for a process to be projected to the achievement of high goals through a continuous reorganization of its internal structures, based on the permanent changing of the dependent relationship with the context. The change in the relations of dependence favors the internal reorganization through feedback. For example a teacher can tackle the learning process of a numerical competence from a self- organization frame in the following way:

1. To orient the students to determine their learning goals in relation to the numerical competence in mathematics in the ninth grade of basic education.
2. To guide the students to transcend their goals and to include major goals in relation to their learning, such as to contribute for a better relationship in the group and in the institution and to contribute to the solution of the problems of humanity.
3. To facilitate situations so that the students reach in a progressive form new understanding in the applicability of the numerical competence from the processes of autonomous search.
4. To orient the students so that they change their dependence relationships with the contexts and to make them include other referents in the learning of mathematics such as classmates, professionals, videos, speeches of experts who apply mathematics in their profession etc.

To search the students to have a continuous feedback from their self-reflex ion and the valuation assessment of other people that favor their continuous learning from a frame of changing relations with the context.

This is a simple example of the application of self- organizing ability that consists synthetically in a process to reach the autonomy and to get goals, from a frame of changing relationships with the context.

What is the knowledge society?

The society is going from the industrial era to the knowledge era, and this is a reality all over the world independently from the cultural or economical conditions. Therefore the knowledge has begun to occupy the leading rol in the production ways. Such a situation is changing the social relationship, communication and the way of living of people. The paper of education is to achieve that the citizens, no matter their economic or social situation, have access to knowledge and to use it for a better quality of life and entrepreneurship.

Industrial society is organized under the capital work, the production in series, industrial equipment and the personal that make some well programmed labour. In this context Frederick Taylor's contributions (1911) were quite relevant in this context. He established the scientific administration by transferring physical sciences paradigm to the management field. The current analysis of such perspective shows us that:

a) The whole is given by the understanding of the different parts.
b) The relationships are given by a lineal causality in which there are causes and effects.
c) The management is to predict to administer and to control.

These proposals have had a notorious impact in the organizational management and in industrialization in the correct emergency of the knowledge society what matters is not to produce goods and services but to generate knowledge that allows permanent creativity and innovation in those goods and services in a more dynamic and interconnected world. In this form, Taylor´s classic epistemology and the approach of scientific administration are not the most pertinent ones. In the knowledge society it is required to perform with flexibility, to accept the continuous change and to tackle diversity. In other words it is to live with a complex way of thinking.

It is necessary that people get trained in knowing how to search, understand, systematize, analyze, create and apply knowledge to talk about everyday situations and to solve problems that come up. It is not a matter of students to have many contents in mind, because knowledge is in a continuous change, but knowing how to process them with a critical attitude, to innovate them and to apply them with pertinence.

Which are the models of quality management?

Another line of work that has done (and it still is doing) essential contributions to the competencies is quality management, in which there are specific models such as the European Foundation Model for quality (EFQM) and the ISO Norm Model. In these two models, the importance of considering factures and dimensions related to human talent is outlined. This is how they have arrived to the incorporation of the competencies approach as a base of quality management in the organizations. For example in the ISO 9001: 2000 standard it is established that the organization must determine which competencies people must have, according the impact of the quality in the achievement of goals. The reason for this is that competencies bring methodological elements so that the organizations have more pertinent human talent with its goals.

Additionally, quality management models have enriched the competencies model, and they are contributing to its positioning not only in the social and entrepreneurial organizations, but also in the educative institutions including universities.

This has gotten competencies to be conceptualized as an approach to the service of quality management, as well as human talent and the learning processes (see Tobón et al.; 2006) because they offer concrete standards of evaluation of these aspects. That is why people have begun to construct models of quality with an explicit integration of the competencies approach as it happens in the Ges FOC model (Tobón; 2013a) .

METHODOLOGY TO TACKLE Competencies

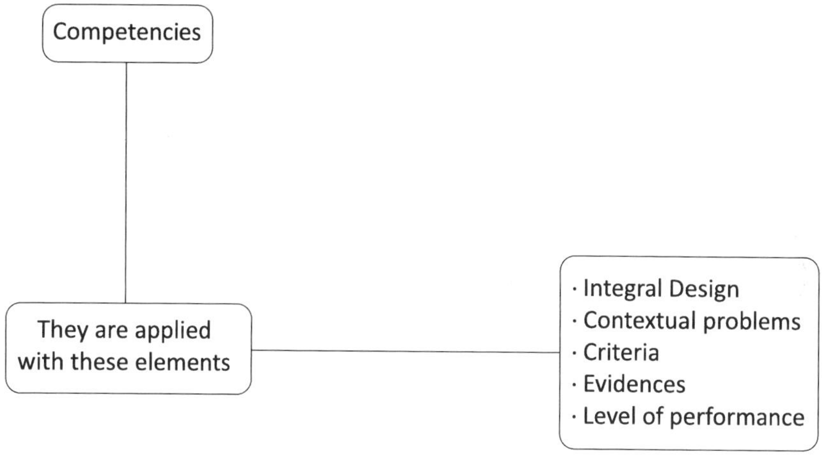

What consists the socioformative methodology of competence formulation on?

There are diverse methodologies to draw up competencies in curriculum and in the formative projects, such as the functionalist behavioral, constructivist and socioformative methodology. The three first are characterized by a great emphasis on the formal aspects and by fragmenting the human performance in multiple components. In the socioformative methodology on the order hand it is characterized for tackling performance in its integrality from contextualized problems with a high grade of flexibility.

In the socioformative methodology it is proposed to work the competencies with the following key axis:

- To establish problems of the context
- To determine competencies
- To build up key criteria
- To set out the necessary and essential evidences
- To formulate the levels of performance

Next, the advantages and characteristics of this methodology are shown as it is applied nowadays in most Latin-American countries.

1. The socioformative methodology is simple: This is because it tackles the central aspects of a competence such as problems of the context criteria and evidences; without including other components that make the process of professors difficult, as the units of competence the elements of competence, fields of application (ranks of development), contents by dimensions (cognitive contents, procedural and affective- emotional) the types of criteria (indicators,, results of learning, attributes and performances), the types of evidences (of knowledge, product of being) etc. These elements complicate the educative labor a lot and they are an obstacle so that professors transform their practices under essential axis.

2. In the socioformation there is a tendency for application, it does not tend to fill up formats. In the traditional methodologies of competencies (functionalist behaviorist or constructivist) there is a great emphasis on planning by detail the learning teaching and evaluation process that are completed later on with a diversity of formats by following, many times, the same teaching postulate, programed in the sixties.

The socioformation goes directly from the studies plan to the practice in the classroom without any detailed planning, because it searches the projects to be agreed and built with the students. For example, in the functionalist methodology, once the curriculum exists, there must be planning for each module in units of learning, and each one of this must contain: learning to be achieved, knowledge and contents in different dimensions, methodology, strategies, activities, sceneries of evaluation etc.

Many teachers do not have the necessary formation to tackle these elements appropriately and, if they have the formation, they do not do it in detail or, because it separates them of the essential; to achieve students to be competent. In the socioformation, on the other hand, from the studies plan, it's beginning to work with the students without any need of planning by detail modules or subjects. We only agree the key activities to be done in the frame of methodology of formative projects. In these, the following axes are worked: competencies to be formed, the contextual problem to be solved. The key criteria to be achieved, the evidences that students must give, the levels of performance to be verified and the activities that will be done to resolve the problem, to fulfill with the criteria and to give the evidences.

3. Contents are a means, not an end. Although it is clear that every day competencies represent a new paradigm in the learning-teaching and evaluation process, -what implies to stop tackling contents as the essence of formation-, the traditional and predominant methodologies in the worldwide scale, still continue a central paper to contents, getting to a point where there is not a possibility of planning learning without describing them by detail. This is making a change to be difficult in the teaching practices because teachers assume that this is the only line to be followed, and they look for contents, in their students´ learning rather than looking for competencies. In the socioformation, contents have been emerged and everything is focused in solving problems of the context, by trying students to learn strategies to search, process, analyze, criticize, create, adapts, innovate and apply the contents, without having to teach them.

What components does the socioformative methodology have?

As it is illustrated in the axis of exemplification here after, the five components of the socioformative methodology are described, with some suggestions about its writing. These elements are only a referent. Each institution can adapt them in its particularities and educative model.

1. Writing of the problems of the context

In the traditional approaches, competencies tend to be tackled based on tasks. In the socioformation, on its side, they are worked through contextualized problems; which are identified from documental review, interviews to experts and in field studies present and future. (frame 1.11).

Frame 1.11 Elements to tackle problems.

Types of problems	· A non answer need in the context · Little knowledge about a phenomenon · Contradiction between two or among more approaches, theories and methodologies. · Need to improve something. · Challenge of creation · Challenge of innovation. Etcetera.
Writing	· In question form · In affirmative form
Identification of problems	To follow up this order: · To review the existent documentation about studies of the context · To interview to experts · To do field studies

2. Writing of the competencies

Competencies are established by considering contextual problems and they can be written in different ways. Although the socioformation proposes the elements to be used: a verb conjugated in present tense, a conceptual content, purpose and a condition of the context. (frame 1.12).

Frame 1.12 Guidelines to write down competencies

Verb conjugated in present form.	A verb in present tense is suggested for the following reasons: a single verb facilitates having a clear idea of the performance, to focus on its formation and to evaluate it. When there are several verbs, this makes it more difficult and the students lose their goal. On the other hand, the verb in present reaffirms that competencies have to be worked on here and now, they do not have to be worked as goals for the future, as it happens with objectives. However, if the institution agrees to work on verbs in infinitive form, or to have several verbs on a competence, it can also be accepted by the socioformation, because the most important thing is to achieve integral formation.	Example: self-manage
Conceptual content	The conceptual content is a concept, process or ample theory, because it can only remain as a skill on other hand. For example, if the performance verb is "builds", the conceptual content cannot be "doors", because it would remain as: "to build doors", and that is just an ability. Therefore, to have a competence, the conceptual content should be "flats", and in such way, it would remain as a competence: "to build flats"	Example: formation
Purpose	This is the "what for" of the competence. The absence of the purposes is one of the major lacks of the traditional approaches of competencies, because of their emphasis on the behavior and in the tasks. In the socioformation, because of the following up of complex thinking, its purpose is planned to be described in each competence. An alternative option is to describe the purposes in the educative model, on condition that the professors and students have full command of it and that they have it in mind in all the formative projects.	Example: to reach the personal fulfillment and the established goals.
Condition of the context	They are the general conditions that must be fulfilled in the competence. These conditions help to understand the competence in its integral form. They constitute an implicit must be. From there, its relationship with the know how to be, and ethics.	Example: with responsibility by facing the challenges of context and by applying meta cognition.
Self formation competence: to self-manage the self-formation throughout life to reach personal fulfillment and established goals, with responsibility; by facing the challenges of context and by applying meta cognition.		

However, competencies can also be written in a free form, on condition that the performance is clear, as an integral performance. In this case, it does not matter the ending of the verb, or the sequence of the components. To be able to elect this last form of writing, it is necessary to have a higher rank of appropriation of the concept of competencies among all the actors of the formation process and to make sure that the appointing agency of educative quality accepts this proposal.

In the writing of competencies there must be used verbs of performance. In the frame 1.13 are some fundamental verbs that must be used in the description and writing of competencies.

Frame 1.13 examples of performance verbs to write down the competencies (Tobón, 2012 a)

Performance verbs for competencies of medium complexity	*Examples of competencies of medium complexity*	*Performance verbs for competencies of high complexity*	*Examples of competencies of high complexity.*
To agree, to cheer, to formulate, to help, to attend, to support, to search, to codify, to prove, to communicate, to build, to consult, to control, to talk about, to show, to say, to dictaminate, to design, to disclose, to elect, to employ, to express, to induce, to inform, to locate, to handle, to manipulate, to drive, to observe, to operate, to prioritize, to pick up, to recuperate, to register, to register, to resolve, to systematize, to transmit, to use, to utilize.	· To Operate programs of office automation to fulfill with activities of processing of information, based on determined organizational goals. · To register information so that the organization can take bases decisions in reliable facts, by considering a determined system of captured data of the processes.	To accompany, to advise, to adapt, to administer, to consult, to audit, to train, to comment, to create, to develop, to diagnostic, to direct, to run, to discuss, to edify, to execute, to undertake, to teach, to evaluate, to explain, to make, to form, to process, to innovate, to intervene, to investigate, to improve, to negotiate, to organize, to guide, to plan, to prevent, to produce, to program, to propose, to protect, to project, to publish, to reenact, to represent, to feedback, to socialize, to transform, to link.	To administer the organizational processes to achieve the vision, mission and strategic objectives of the company, by having in mind the quality assurance. To evaluate the execution of the organizational processes to make convenient and pertinent decisions that lead to achievement of goals, with precedent information of different sources of the company.

3. Writing of Criteria

The criteria are performances or concrete results that orient to formation as well as evaluation of competencies. It is suggested to have among three or eight criteria for each competence (frame 1.14).

Frame 1.14 Suggestions to write down the criteria.

Number	There must be two criteria at least. Regarding the maximum number of criteria by competencies there is not any limited guidelines , but it is recommended not to have more than 10. To have among four and eight criteria helps for mediation and valuation a lot.
Writing	The writing of criteria is in the same form that the one of competences, but without the finality. This means that each criteria includes: · A verb in present tense. · A concrete conceptual content · Condition of the context.
Methodology	To determine the criteria these recommendations must be followed: 1.- to review the problems of the context and the competence or the competences. 2.- to avoid the adjectives in the writing, such as excellent, very good, etcetera. 3.- to look forward criteria to be evaluable. 4.- to make sure the criteria to tackle key and relevant aspects of the competence.

On the frame 1.15 there are some examples of verbs to write down the criteria. 10 minimal processes are considered that a competence implies, according to the socioformative theory in Tobón (2013 a). Such processes are:

1.- Sensitization
2.- Conceptualization
3.- Solution of the problems
4.- Values and ethical project of life
5.- Collaboration with others
6.- Assertive communication
7.- Creativity, personalization and innovation
8.- Transverse and transference
9. Management of resources
10. Assessment

Frame 1.15

Number of process	*Competence process*	*Competence process*	*Examples of general criteria*	*Examples of concrete criteria*
1	Sensitization	To admire, to deal with, to codify, to enjoy, to watch, to perceive, to prefer, to process, to want, to concentrate to get interested in, to motivate, to finish, to visualize.	To motivate about fulfillment of activities. and/or problem solving, keeping established goals in mind. To concentrate in problem analysis, according to expected goals. To finish the activities that begin or the problems that tackle according to what is planned.	To have motivation about the study of media when developing values, which is shown in the depth of discussion taken.
2	Conceptualization	To look for (information), to characterize, to categorize, to classify, to compare, to conceptualize, to define, to determine, to differentiate, to exemplify, to identify, to enquire, to justify (the concept), to process (information), to recuperate, to relate, to subdivide, to systematize.	To define with your own words the necessary basic concepts, to work on situations and problems of the area, by considering and determining bibliographic sources. To characterize the essential concepts of the area, by keeping in mind bibliographic sources of the last three years. To justify the importance of the concept to work on a specific situation or problem, by considering at least two bibliographic sources of the last three years. To differentiate from other closely similar concepts which are in the same category or those they can be confused with , by using bibliographic sources.	To define with your own words the concept of ecosystem by considering a specific work published in the last three years, that can be used as a reference

3	Solution of the problems (interpretation, argument, and solution)	Sub process: analysis. To Analyze, to argue, to understand, to show, to diagnose, to explain, to interpret, to justify, (the problem), to summarize, to synthetize. Subprocess: resolution To apply, to build, to control, to develop, to perform, to design, to edify, to make effect, to execute, to elaborate, to start, to use, to undertake, to make, to form, to do, to implement, t o elaborate, to handle, to manipulate, to improve, to operate, to plan, to produce, to program, to propose, to project, to pick up, to register, to represent, (a situation or a problem), to resolve, to utilize, to wear.	· To diagnose the situations and problems according to a determined and pertinent procedure of the area. · To interpret the situations or problems with concepts or pertinent theories to the area, which have a back up in a bibliographic source at least. · To explain the causes and consequences of a determined problem, by considering the context in which it appears and the antecedents. · To justify the importance of analyzing the problem, and to consider determined facts and referents of academical bibliography. · To propose solutions to the problems by considering the diagnostic of base, and taking into account among other elements, resources, the required human talent and the time. · To plan the tackling of the situations and the problems, To consider a determined diagnostic and pertinent procedures in the area, with an adequate base line. · To resolve the situations and problems by following some established and base lined procedures, by considering he goals too. · To produce solutions to the situations and problems that correspond with the diagnostic and the expected goals. · To register the process of resolution of the situations and problems by having as a base a determined, pertinent methodology in the context.	To diagnose the most common types of violence in the family, having as a base a pertinent instrument of the area. To explain the causes and consequences of violence in the family, by considering diverse, personal, economical or cultural factors. To propose a solution of the problem of obesity in youngsters between 12 and 18 years old, by considering the availability of resources, the comfort resources and the necessary time that must be dedicated. To resolve a problem of an inadequate nutrition in a person, by considering the previous diagnostic and the available resources. To register the process of continuous improvement of the diet, with base on the standards of a healthy nutrition.

4	Values and ethical project of life	To assume, to correct, to take care of, to fulfill, to honor, to behave, to practice (values), to prevent, to protect, to acknowledge, to keep, to respect, to answer, to have self- esteem, to commit.	To fulfill with the activities according to a specific programming. To respect the basic standards in order to work on a situation or problem, taking into account laws and human rights. To answer about the consequences of his acts, by preventing and fixing the possible errors in little time and with minimal, secondary consequences, according to the universal values and the laws. To act honestly in diverse situations and problems, according to the challenges of the context.	To be responsible in the delivery of information of the solved problems by applying Mathematics, according to an established programming. To respect the standards of collaborative work in the process of problem solving with mathematics.
5	Collaboration with others	To accompany, to advice, to agree, to cheer, to consult, to help, to assist, to train, to collaborate, to share, to complement, to coordinate, to direct, to leader, to motivate, to negotiate, to orient, to participate, to challenge, to integrate, to socialize, to bear, to have guardianship of.	To agree with other people the fulfillment of determined activities and the fulfillment of concrete goals, considering a need or a problem to be solved. To collaborate with other people in the resolution of the problems, according to a determined planning and the established goals. To participate in collaborative processes by doing the activities that correspond to the person, according to a determined planning To complement the actions of others in the collaborative processes, by considering the planning and goals that are looked for.	To collaborate with his peers in the fulfillment of a project to apply the technology in the improvement of quality life, according to the availability of resources and determined goals. To complete the wiki done by classmates or teamwork about the positive and negative effects of globalization, with the objective of explaining in a more detailed form this phenomenon of global society.

6	Assertive communication	To announce, to advise, to comment, to communicate, to talk about, to say, to speak about, to publish, to write, to listen to, to express, to notify, to divulge, to feedback.	To listen to others according to a specific goal and by regulating one's attention. To communicate different ideas and feelings with clarity and warm, by respecting others rights, ideas and emotions. To act by following assertive expressions, in relation to determined goals. To feedback others by focusing in his achievements and aspects to be improved, with clarity. To respect ideas and emotions, based on common referents.	To listen to with attention to his team mates during the rehearsal of a play about drugs intake which allows the achievement of expected goals. Saying "no" to drug offers in the community, by showing an assertive behavior and acting on the consequence.
7	Creativity, personalization, and innovation.	To accommodate, to adapt, to create, to innovate, to make personal, to recreate.	To create solutions for the problems by considering the management of the resources and indictors of effectiveness, efficacy and efficiency. To adapt a theoretical model to the explanation of a situation or problem, by considering diverse factors that influence in an evolutionary frame. To personalize the attainment of a procedure considering to set goals. To innovate procedures that can solve specific problems from clear diagnosis and with precise arguments of achievements.	To create a solution of environmental management to prevent global warming from home. This allows a saving in natural and economical resources. To adapt prevention strategies for global warming into the family context and to consider the home context and specific cultural factors.

8	Transversality and transference	To harmonize, to articulate, to teach (to others), to pass, to relate, to work, (from the multi, inter or transdisciplinariety) To transfer, to transversalize, to transmit, to trespass, to join.	To articulate knowledge of different areas or disciplines when tackling situations and problems, by looking for complementarity of such knowledge. To work in inter disciplinary activities and problems, based on concrete goals and achieving better processes and results than in disciplinary tackling. To transfer determined procedures from one area to other, based on searched goals .	To solve domestic or family economy problems by applying Mathematics , the principles of economy and the opportunities of purchase of products in the context. To propose an integral solution to an economical problem by considering the theoretical and methodological contributions of at least, two different disciplines of economy, looking for the complementation of contributions.
9	Management of resources	To prepare , to adapt, (resources), to adjust, to mold, to look for (resources), to frame, to scrutinize, to explore, to manage, (resources), to organize, (to have), RECURSIVITY	To look for necessary resources for the procedures and the achievement of goals, within the context. To set the resources to the activities and goals, by having in mind the possibilities of the context. To adapt technological equipment to expected processes and goals thanks to the identified needs.	To look for necessary resources to resolve a historical problem of relevance for the real context, by considering the possibilities of the context. To create a mechanism to organize historical information about a problem, keeping expected goals in mind.

10	Assessment	To self-evaluate, to examine, to self regulate, to value, to prove, to-value, to evaluate, to heteroevaluate, to inter-value, to meta-evaluate.	⊠ Self evaluate his information in a continuous manner, pointing out his achievements and aspects to be improved, by implementing concrete actions of improvement in the context, according to concrete goals. ⊠ Covalue the performance of other people, by taking as a base determined criteria and instruments. ⊠ Meta evaluate the processes and results of evaluation with base on certain criteria.	To Evaluate the quality of public service of water supply in a locality by applying determined principles and laws of sciences and considering the context. Self- regulate his performance in the public service of water supply in the locality diagnosis, considering the established goals .

Writing of evidences

Evidences are concrete products that students must present to show the achievement of criteria, and the formation of the expected competence. Several competencies can be evaluated with the same evidences. In the frame 1.16 are some suggestions shown to write them down.

Frame 1.16- suggestions to write down evidences.

Number	We can have only one integrative evidence or more than that. From socioformation, it is recommended to have few evidences and to make them fundamental. This facilitates the assessment, the following up and the continuous improvement.
Writing	The product is written first in the evidences, and it´s followed up by a little description. Examples: · Register of oral exhibition about stomach cancer prevention. · Report of the improvement project of rural housing. · Video of teamwork · Essay about the actions to prevent violence. · Written test about strategies to avoid drug consumption.
Methodology	The product is written first in the evidences, and it´s followed up by a little description. Examples: · Register of oral exhibition about stomach cancer prevention. · Report of the improvement project of rural housing. · Video of teamwork · Essay about the actions to prevent violence. · Written test about strategies to avoid drug consumption.

On frame 1.17 there is a proposal of a format to write down a competence like an integral performance. This contrasts with the emphasis in tasks and skills of the traditional approaches that must be overcome in the knowledge society.

Frame 1.17 format of description of a competence in the socioformative approach.

Competence	
Problem of the context	
Criteria	***Evidences***

Levels of performance

The levels of knowledge or performance are the different grades or phases in which competencies are developed. In the competencies, there are different theories of levels, one of them is the socioformative one, which proposes five levels, which go, from pre-formal until strategic level. It is essential to have the five levels of knowledge to mediate the formation in projects and to value the competencies, as well as to ensure the continuous and progressive improvement of the students.

The levels of knowledge are worked with different instruments of evaluation, one of them is the learning map, as it can be appreciated in the example that was presented at the beginning of this chapter (frame 1.2). In this example, in each level of performance the achievement for each criteria is determined, with the objective of establishing how the person is, and which actions of help must be implemented so that this person achieves, as minimum, the resolute level, and then , he or she advances to the strategic level.

In the learning maps, the quantitative deliberation can be added, over all if it is necessary the student to have a note for the follow up and accreditation of the studies. This deliberation is basically determined according to the following guidelines:

- Some points or percentages are determined for the evidence, according to the grade of relevance to show the considered criterion or criteria.
- The points or percentages of the evidence are distributed among the describing elements or indicators of the different levels of performance.
- The resolute level is given the half or more than the half of points or percentages are given.
- The other levels are given points or percentages according to their grade or challenge. To do this, it is well known that the higher the grade of analysis, argument, creativity and /or complexity (number of considered factors and relations among them) the higher quantitative deliberation will be.

In the frame 1.18, some essential characteristics of each one of the proposed levels for a competence are described. It is necessary to point out that a superior level includes the characteristics of the previous levels. For basic education and medium education it is enough to have four levels of performance: receptive, resolute, autonomous, and strategic, which helps more to understand of the process by the students.

Frame 1.18 Levels of performance of the competencies.

Level	Characteristics (one or more)	Example: Leadership
Preformal	· There are some general ideas of the competence.	· To have some general ideas about work with other people.
Receptive	· There is some concrete knowledge about competencies. · There is reception and understanding of the information. · Actions are sometimes too operative. · There is motivation although it is not constant.	· The person sometimes formulates suggestions of how to improve and to go on with the activities · This person sometimes considers the vision and vision of the organization.
Resolutive	· Simple problems are resolved · There is participation with technical elements. · Very concrete procedures are followed · There are basic concepts	· To contribute to keep a motivated team about goals achievement, by having in mind the purposes of the area. · Help to resolve problems in the team to reach the goals.
Autonomous	· There is criterion and initiative · There is a resolution of the problems that imply several variables. · There is a management of processes. · There is scientific argument. · There is direction of projects.	· To direct projects and succeed with them · He is a referent of a leader in his area. · Promote the mission and vision of the organization in the different activities.
Strategic	· Strategies of change in the context are applied. · There is a high flexibility to do things. · There is a high grade of innovation. · There is a high commitment with humanity	· To generate commitment and motivation in the members of the organization for new goals. · He is an institutional referent of commitment and professionalism.

In a curriculum of competencies it is required to value the grade of formation of these in the students, through the implementation of didactic, pertinent strategies. This is key so that they do not stay in paper, but to contribute to the formation of people with a solid, ethical project of life and entrepreneurship. In the frame 1.19, a proposal of format of a learning map is proposed, to orient the evaluation of competencies in the studies plan. This instrument can be applied in the evaluation of diagnostic, in the continuous evaluation (follow up evaluation), or in the summative evaluation (accreditation evaluation), in the modules or projects that exist.

Bibliographic references

Alles, M. (2005a) *Desempeño por competencias [Performance by competencies]*. Buenos Aires: Granica.

Alles, M. (2005b). *Diccionario de comportamientos. Gestión por competencias [Dictionary of behaviors. Management by competencies]*. Buenos Aires: Granica.

Arthur, W. B. (1995). *Increasing Returns and Path Dependence in the Economy.* Ann Arbor: University of Michigan Press.

Barbero, J.M. (2004). *Saberes hoy: diseminaciones, competencias y transversalidades [Knowledge today: disseminations, competencies and transversalities].* Revista Iberoamericana de Educación, No. 32.

Barrón, C. (2000) *La educación basada en competencias en el marco de los procesos de globalización [Competence-based education within the framework of globalization processes]*. En M. Valle (coord.), *Formación en competencias y certificación profesional [Training in competencies and professional certification.]*. México: Centro de Estudios sobre la Universidad. UNAM.

Begun, J.W., Dooley, K., y Zimmerman, B. (2003). *Health care organizations as complex adaptive systems.* En: Mick, S. M., y Wyttenbach, M. (Eds.), *Advances in Health Care Organization Theory.* San Francisco: Jossey-Bass.

Bloom, B.S., et al. (1956). *Taxonomy of educational objectives, cognitive domain.* New York: Longmans, Green.

Bogoya, D. (2000). *Una prueba de evaluación de competencias académicas como proyecto [A test of academic competencies evaluation as a project]*. En: Bogoya, D. y colaboradores. *Competencias y proyecto pedagógico [Competencies and pedagogical project]*. Santafé de Bogotá, D. C: Unibiblos.

Boyatzis, R. (1982). *The competent manager*. New York: Wiley and Sons.

Brunner, J. (1992). *Actos de significado. Más allá de la revolución cognitiva [Acts of meaning. Beyond the cognitive revolution]*. Madrid: Alianza.

Brunner, J. (1997). *La educación, puerta de la cultura [Education, gate for culture]*. Madrid: Visor.

Bustamante, G. (2003). *El concepto de competencia III. Un caso de recontextualización: Las "competencias" en la educación colombiana [The concept of competence III. A case of recontextualization: The "competencies" in colombian education]*. Bogotá: Sociedad Colombiana de Pedagogía.

Cariola, L. y Quiroz, A. (1997). *Competencias Generales, competencias laborales y currículum [General Competencies, labor competencies and curriculum]*. En Novick, M. y Gallart, M. A. (coords.), *Competitividad, redes productivas y competencias laborales [Competitiveness, productive networks and labor competencies]* (pp. 51-72). Montevideo: CINTERFOR.

Chomsky, N. (1970). *Aspectos de la teoría de la sintaxis [Aspects of syntax theory].* Madrid: Editorial Aguilar.

Condemarín, M. y Medina, A. (2000). *Evaluación auténtica de los aprendizajes. Un medio para mejorar las competencias en el lenguaje y comunicación [Authentic evaluation of learnings. A way to improve competencies in language and communication].* Santiago de Chile: Editorial Andrés Bello.

Conferencia Mundial para la educación superior (1998). *La educación superior en el siglo XXI: Visión y acción [Higher education in 21st century: Vision and action].* Paris, Octubre 5-9.

Domínguez, E. (2006). *Pensamiento complejo y educación. Una síntesis de las tesis de Morin [Complex thought and education. A synthesis of Morin's theses].* En S. Tobón et al. (Eds.), *Diseño curricular por competencias [Curricular design by competencies].* Medellín: Uniciencia.

Echeverría G., M. (2004). *¿Valores o competencias? Dos enfoques formativos en las prácticas pedagógicas. Informe área de educación, 10 [Values or competencies? Two formative approaches in pedagogical practices. Education Area Report, 10].* Universidad Santo Tomás Santiago de Chile, julio 2004.

Escohotado, A. (1999). *Caos y Orden [Chaos and Order*]. Madrid: Espasa Calpe.

Flavell, J.H. (1987). *Speculations about the nature and development of metacognition.* En F.E. Weinert y R.H. Kluwe (Eds.), *Metacognition, motivation and understanding.* Hillsdale, NJ: Erlbaum.

Gallego, R. (2000). *El problema de las competencias cognoscitivas, una discusión necesaria [The problem of cognitive skills, a necessary discussion].* Santafé de Bogotá D. C.: Universidad Pedagógica Nacional.

Gardner, H. (1987). *Las estructuras de la mente. La teoría de las inteligencias múltiples [Mind structures. The theory of multiple intelligences].* México: Fondo de Cultura Económica.

Gardner, H. (1995). *Inteligencias múltiples. La teoría en la práctica [Multiple intelligences. Theory in praxis].* Barcelona: Paidós.

Gardner, H. (1997). *La mente no escolarizada. Cómo piensan y cómo deberían enseñar en las escuelas [The non-schooled mind. How schools think and how they should teach.].* México: SEP/Coop. Española, Fondo Mixto de Coop. Técnica y Científica México-España.

Gell-Mann, M. (1995). *El Quark y el jaguar: aventuras en lo simple y lo complejo [The Quark and the jaguar: adventures into the simple and complex].* Barcelona: Tusquets Editores.

Goleman, D. (1996) *La Inteligencia Emocional [The Emotional Intelligence].* Buenos Aires: Vergara.

González, J., y Wagenaar, R. (2003). *Tuning educational structures in Europe.Informe final fase uno.* Bilbao: Universidad de Deusto.

Hernández, C.A., Rocha, A., y Verano, L. (1998). *Exámenes de Estado. Una propuesta de evaluación por competencias [State Exams. A proposal for evaluation by competencies*]. Bogotá: ICFES.

Holland, J. H. (1996). *Hidden order: how adaptation builds complexity.* Cambridge: Perseus Books.

Holland, J. H. (1998). *Emergence: from chaos to order.* Oxford: Oxford University Press.

Hymes, D. (1996). *Acerca de la competencia comunicativa. Forma y Función, 9 [About communicative competence. Form and Function, 9].* Departamento de Lingüística, Universidad Nacional de Bogotá.

IBERPROF (2000). *Análisis ocupacional y funcional del trabajo [Ocupational and functional analysis of work].* OEI.

Kauffman, S. A. (1993). *Origins of order: self-organization and selection in evolution.* Oxford: Oxford University Press.

Kauffman, S. A. (1995). *At home in the universe: the search for the laws of selforganization and complexity.* New York: Oxford University Press.

Kuhn, T. (1962). *La estructura de las revoluciones científicas [The structure of scientific revolutions].* México: Fondo de Cultura Económica.

Levy-Leboyer, C. (1997). *Gestión de las competencias [Management of competencies].* Barcelona: Gestión 2000.

Lyotard, J.F. (1993). *La condición posmoderna [The postmodern condition].* Barcelona: Planeta Agostini.

Massot y Feisthammel (2003). Seguimiento de la competencia y de la formación. Madrid: Ediciones AENOR.

Maturana, H., y Varela, F. (1994). *El árbol de conocimiento [The tree of knowledge].* Santiago de Chile: Editorial Universitaria.

Max-Neef, M. (1993). *Cultura, economía y diversidad en el mundo actual [Culture, economy and diversity in today's world].* En *Aprender para el futuro: Educación ambiental [Learning for the future: Environmental education]*. Documentos de un debate. Fundación Santillana.

Max-Neef, M. (1996). *Desarrollo a escala humana [Development on a human scale].* Santiago de Chile: Fundación CEPAUR.

McClelland, D.C. (1973). *Testing for competencies rather than intelligence.* American Psychologist, 28, 1-14

McClelland, D.C. (1993). *Introduction.* En L.M Spencer y S.M. Spencer (eds.), *Competence at work.* New York: John Wiley and Sons.

Morin, E. (1995). *Introducción al pensamiento complejo [Introduction to complex thought].* Barcelona: Gedisa.

Morin, E. (2000a). *Presentación [Presentation].* En *Memorias del Primer Congreso Internacional de Pensamiento Complejo [Memories of the First International Congress of Complex Thought]*. Bogotá: Ministerio de Educación Nacional.

Morin, E. (2000b). *La mente bien ordenada [The well-ordered mind].* Barcelona: Seix Barral.

Morin, E. (2002a). *Los siete saberes necesarios para la educación del futuro [The seven knowledges required for the future education].* Buenos Aires: Nueva Visión.

Morin, E. (2002b). *Ética y globalización [Ethics and globalization].* En *Los Desafíos Éticos del Desarrollo [The Ethical Challenges in Development].* Buenos Aires: 5 y 6 de septiembre.

Orozco, B. (2000). *De lo profesional a la formación en competencias [From professional to the formation in competencies].* En M. Valle (coord.), *Formación en competencias y certificación profesional [Formation in competencies and professional certification].* México: Centro de Estudios Sobre la Universidad. UNAM.

Ortega y Gasset, J. (2004). *Meditación de la técnica y otros ensayos sobre ciencia y filosofía [Meditation of technique and other essays about science and philosophy].* Madrid: Alianza.

Pfeffer, J. (1999). *La ecuación humana [The human equation].* AEDIPE. Madrid: Ediciones Gestión-2000.

Prigogine, Ilya (1997). *The end of certainty: time, chaos, and the new laws of nature.* New York: The Free Press.

Prigogine, Ilya, y Stengers, Isabelle (1984). *Order out of chaos: man's new dialogue with nature.* New York: Bentam.

Senge, P. (1995). *La Quinta Disciplina en la práctica [The Fifth Discipline in practice].*

Buenos Aires: Granica.
Spencer, L.M. y Spencer, S.M. (1993). *Competence at Work.* New York: John Wiley and Sons.
Sternberg, R.J. (1997). *Successful intelligence.* New York: Simon and Shuster.
Taylor, F. W. (1911). *The principles of scientific management.* New York: Harper & Brothers.
Tobón, S. (2004). *Cartografía conceptual [Conceptual cartography].* Islas Baleares (España): CIBER EDUCA.
Tobón, S. (2005). *Formación basada en competencias. Pensamiento complejo, diseño curricular y didáctica [Formation based in competencies. Complex thought, curricular design and didactic]* (2da. Ed.). Bogotá: ECOE.
Tobón, S. (2006). *Las competencias en la educación superior. Políticas hacia la calidad [Competencies in higher education. Policies towards quality].* Bogotá: Ecoe.
Tobón, S. (2009a). La formación humana integral desde el proyecto ético de vida y el enfoque de las competencias [Integral human formation from the ethical life project and the competency approach]. En E. J. Cabrera (Ed.), *Las competencias en educación básica: un cambio hacia la reforma [Competencies in basic education: a change towards the reform].* México: Secretaría de Educación Pública.
Tobón, S. (2009b). *Proyectos formativos: didáctica y evaluación de competencias [Formative projects: didactics and evaluation of competenecies].* En E. J. Cabrera (Ed.), *Las competencias en educación básica: un cambio hacia la reforma [Competencies in basic education: a change towards the reform].* México: Secretaría de Educación Pública.
Tobón, S. (2010). *Proyectos formativos: metodología para el desarrollo y evaluación de competencias [Formative projects: methodology for development and evaluation of competencies]* México: Book Mart.
Tobón, S. (2011a). *El currículo por competencias desde la socioformación. ¿Cómo podemos cambiar nuestras prácticas educativas para asegurar la formación integral? [The curriculum by competencies from the socio-formation. How can we change our educational practices to ensure integral formation?]* México: Limusa.
Tobón, S. (2011b). *La evaluación de las competencias por medio de mapas de aprendizaje: una propuesta frente a los métodos tradicionales de evaluación [The evaluation of competences through learning maps: a proposal against the traditional methods of evaluation].* En Moya, J. y Luengo, F. (coords.), *Estrategias de cambio para mejorar el currículum escolar: Hacia una guía de desarrollo curricular de las competencias básicas [Strategies of change to improve the school curriculum: Towards a curriculum development guide of the basic competences]* (pp. 135-149). Madrid: Ministerio de Educación. Disponible en: http://www.cife.ws/comunidad
Tobón, S. (2011c). *Manual sintético de gestión del talento humano por competencias y la Quinta Disciplina [Synthetic manual of human talent management by competencies and the Fifth Discipline].* México: CIFE.
Tobón, S. (2012). *El enfoque socioformativo y las competencias: ejes claves para transformar la educación [The socio-educational approach and the competences: key axes to transform education].* En Tobón, S. y Jaik Dipp, A. (Eds.), *Experiencias de aplicación de las competencias en la educación y el mundo organizacional [Experiences applying competencies in education and the organizational world].* México: Redie.
Tobón, S. (2013a). *Formación integral y competencias. Pensamiento complejo, currículo, didáctica y evaluación [Integral formation and competencies. Complex thought, curriculum, didactics and evaluation].* (4ta. Ed.). Bogotá: Ecoe.
Tobón, S. (2013b). *Evaluación de las competencias en la educación básica [Evaluation of*

competencies in basic education]. (2 ed.). México: Santillana.

Tobón, S. (2013c). *Planes y programas de estudio por ciclos propedéuticos y competencias. Hacia un nuevo paradigma curricular desde la socioformación [Plans and study programs by propaedeutic cycles and competencies. Towards a new curricular paradigm from socio-formation]*. México: Trillas.

Tobón, S. y Agudelo, H. (2000). *Pensamiento complejo y formación humana en el sistema educativo colombiano [Complex thought and human formation in the colombian educative system]*. En *Memorias del Primer Congreso Internacional de Pensamiento Complejo, Vol. II [Memories of the First International Congress of Complex Thought, Vol. 2]*. Bogotá: Ministerio de Educación Nacional.

Tobón, S. y García Fraile, J.A. (2006). *La formación de las competencias en la educación superior [The formation of competencies in higher education]*. Madrid: Universidad Complutense de Madrid.

Tobón, S. y Mucharraz, G. (2010). *¿Cómo aplicar las competencias en la práctica docente? [How to apply the competencies in the teaching practice?]*. México: Kundrivi.

Tobón, S. y Núñez, A. (2006). *La gestión del conocimiento desde el pensamiento complejo: un compromiso ético con el desarrollo humano [Knownledge management from the complex thought: an ethic compromise with the human development]*. Revista Escuela de Administración y Negocios, 58, 27-40.

Tobón, S., García Fraile, J.A., Carretero, M. y Rial, A. (2006). *Competencias, calidad y educación superior [Competencies, quality and higher education]*. Bogotá: Magisterio.

Tobón, S., Pimienta, J., y García Fraile, J.A. (2010). *Secuencias didácticas: aprendizaje y evaluación de competencias [Didactical sequencies: learning and evaluation of competencies]*. México: Pearson.

Torrado, M.C. (1995). *La naturaleza cultural de la mente [The cultural nature of mind]*. Bogotá: ICFES.

Torrado, M.C. (1998). *De las aptitudes a las competencias [From aptitudes to competencies]*. Bogotá: ICFES.

Torrado, M.C. (2000). *Educar para el desarrollo de las competencias: Una propuesta para reflexionar [Educating for the development of competencies: A proposal for reflection]*. En D. Bogoya et al. (Eds.), *Competencias y proyecto pedagógico [Competencies and pedagogical project]*. Bogotá: Universidad Nacional.

Torres, E. (2001). *Las competencias, una aproximación desde Aristóteles [Competencies, an Aristotelian approach]*. En: E. Torres, L.F. Marín, G. Bustamante, J.H. Gómez, y E. Barrantes (Eds.), *El concepto de competencia: Una mirada interdisciplinar [The concept of competency: an interdisciplinary insight]*. Bogotá: Sociedad Colombiana de Pedagogía.

Unión Europea (2005). *Recomendación del Parlamento Europeo y del Consejo de la Unión Europea sobre las competencias clave para el aprendizaje permanente [Recommendation of the European Parliament and of the Council of the European Union on key competences for permanent learning*]. http://www.ccoo.es/comunes/temp/recursos/99999/35637.pdf.

Vasco, C.E. (2003). *Objetivos específicos, indicadores de logros y competencias ¿y ahora estándares? [Specific objectives, indicators of achievements and competencies, and now standards?]* Educación y Cultura, 62, 33-41.

Vigotsky, L. (1985). *Pensamiento y lenguaje [Thought and language]*. Buenos Aires: La Pléyade.

Woodruffe, C. (1993). *What is meant by a competency? Leadership and Organization Development Journal,* 14 (1), pp.29-36.

Frame 1.19 Format of lea-ning map. Socioformative rubric (Tobón 2013b)

Program of studies:			**Module or project:**		
Problem of the tackled context:			**Type of evaluation (diagnostic, continuous or summative)**		
Evidence:					
Criterion	**Preformal level**	**Receptive level**	**Resolutive level**	**Autonomous level**	**Strategic level**
1.					
Weighting					
2.					
Weighting					

Program of studies:			Module or project:		
Problem of the tackled context:			Type of evaluation (diagnostic, continuous or summative)		
Evidence:					
Criterion	Preformal level	Receptive level	Resolute level	Autonomous level	Strategic level
3.					
Weighting:					
4.					
Weighting:					

Evaluation process	*Achievements*	*Recommendations*	*Achieved level*
Self-evaluation			
Co-evaluation			
Hetero-evaluation			
Meta-evaluation			

CHAPTER 2

The Curriculum Management by Competencies from the Socioformative Approach

How to change from objectives to integral performances before?

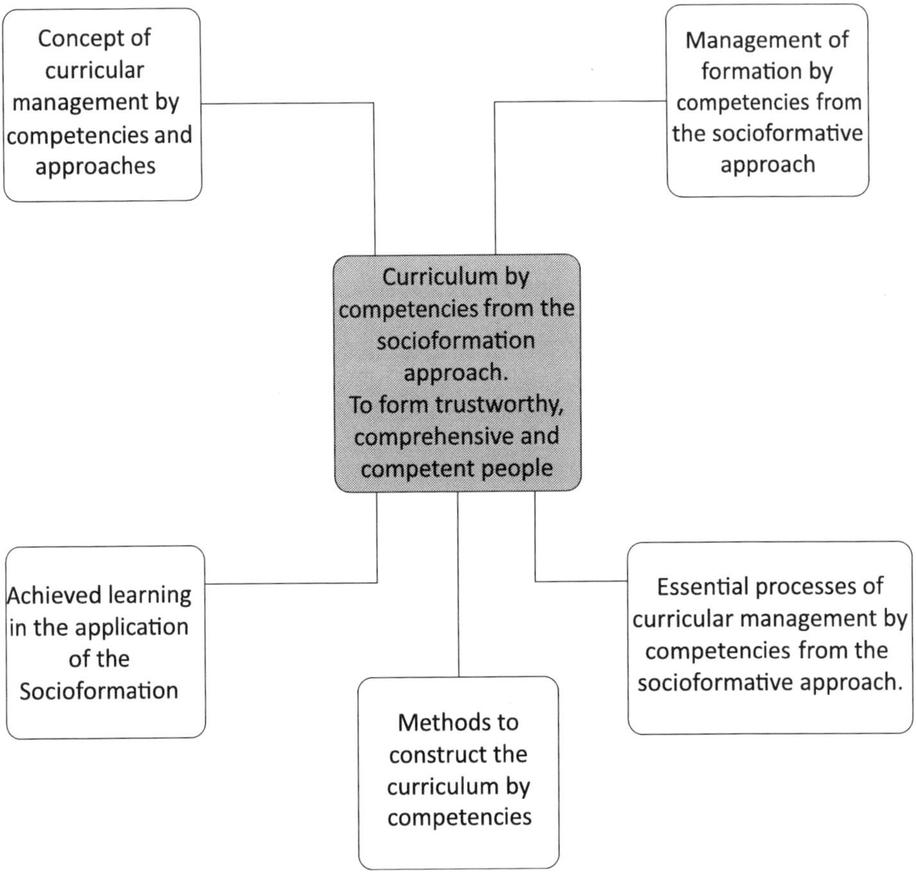

Figure 2.1 Central elements in the curriculum management.

CONCEPT OF CURRICULAR MANAGEMENT BY Competencies AND APPROACHES

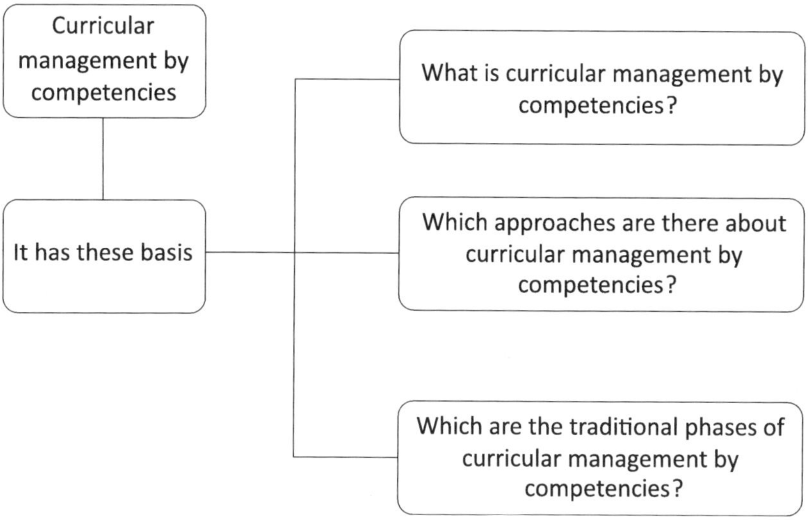

What is the curricular management by competencies?

In general, curricular management consists on implementing plans of formation so that people achieve specific goals, by keeping in mind the institutional philosophy and the external challenges, with the corresponding search and handling of resources, and the necessary human talent; in the historical, social, economic and political frame. This is done in the different types and cycles of education as well as in the organizations.

Curricular management by competencies is part of the curricular management in general and, through the exhaustive study of the context. It intends to determine which basic, generic or specific competencies are going to be set up in the students (graduation profile), through the implementation of a series of formative projects, which are interrelated and organized in the curricular map, with a specific amount of credits, and applying the didactic strategies and evaluation oriented to learning and to the accreditation of the profile competencies, in the learning frame to learn (figure 2.2). In this form, curricular management is a process in a continuous construction, and re-construction, whose goal is to be at the same level of present and future social challenges, so that it can be possible to search the permanent pertinence of formation.

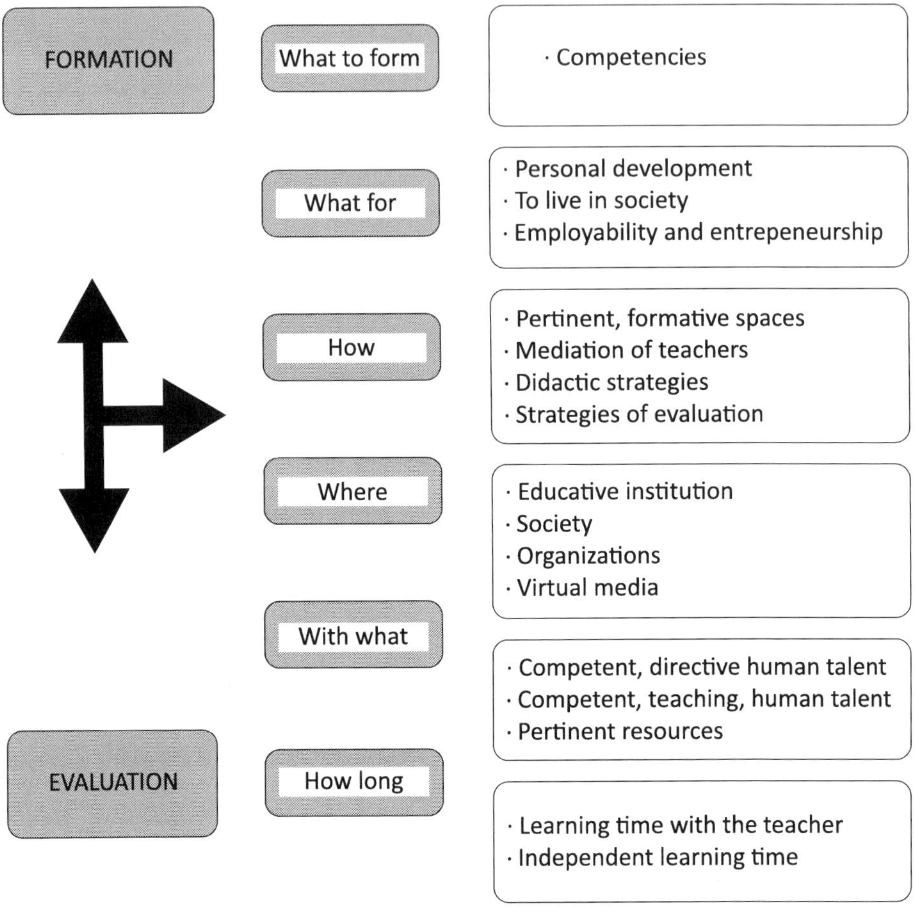

Figure 2.2 The most common aspects of a curriculum by competencies.

Traditionally, the competence approach has been taken in education and in the organizational world as an interrelational process made up by four components: identification of the competencies, normalization of competencies, training based in competencies, and certification of the competencies. However, through practice, there has been a search of agility in the process. Today, the identification of competencies tends to be assumed as part of the normalization. Regarding evaluation and certification they are sometimes conceived as part of the same process, and some other times in a separated form. This happens in Mexico, where there are organizations that evaluate and certify, but also specialized centers in each component. We have decided to leave them separated, to facilitate the analysis and its implementation. In the frame 2.1 are described the main traditional components implied in the implementation of the competencies approach.

Frame.2.1 Essential components of the competencies approach.

1. Standardization of competencies. Establishment of norms of competence to achieve as human talent formation as its certification. In this process, the profiles in the educative world and in the organizational area are also established.	2. Training based in competencies. Desing of the programs of formation of human talent based on standardization of competencies.
3. Evaluation. Determination of the grade in which a person possess determined competence, based on criteria, evidences and instruments of evaluation, with the objective of valuing if it is subject of certification of such a competence, or if he requires of training, complementary activities. (In many cases, evaluation is integrated in the certification of competencies).	4. Certification. Process used by an accredited organization to certify that a person has a determined competence, from the exhaustive evaluation of his development. Previous studies are not required.

What approaches are there about curricular management by competencies?

There are several approaches to manage the curriculum by competencies. With this, it is shown that there is not an only perspective or a single path to follow. When it is intended to have a single perspective, one can fall in the one-direction dimension and there is a tendency to disqualify the work done by other institutions. Therefore, the best is to understand the different approaches, their emphasis and characteristics to have more open and integrative points of view.

The main approaches of curricular management are four: functionalist, behavioural, constructivist and socioformative; all of these with many similarities. For example:

1. They are centered to form competencies.
2. They look formative processes to be pertinent for the students.
3. They take into account several classes of competencies and different knowledge in them.
4. They consider the laboral, social, scientific, cultural changes. Etc.

Next, there is a synthesis of these approaches.

Functionalist approach

This is the most popular and predominant approach in different countries nowadays. It emerged in England through the National Vocational Qualifications, NVQ). As its name expresses, it intends to determine which the functionality of human processes of performance in several contexts is. To achieve this, it is centered in the study of laboral, social and/or scientific functions through the analysis functional method.

In this approach, are studied the activities that people must perform in an area, and based on this, the plan of studies is built, to ensure students to learn such activities. It goes from the external so that it can go later on to the processes of formation. The competencies consist on doing activities and to reach some results in a determined productive function, according to determined criteria of performance.

Behavioral Approach of Curricular Design

This is the result of the historical evolution of the behavioral model and of the educative technology. This is a current approach, and it is based on the development of behaviorism of second and third generations. It consists on orienting the building of the project of formation in an institution from key competencies which give competitive advantages to organizations. This approach is frequently applied in the United States and In Europe.

The methodology of behavioral analysis looks for determining the attributes which increase the success of the workers and professionals through several methods; such as interviews observations of behaviorism in the work, etc. In this manner, some norms of competence are developed, which describe the expected characteristics in people to have a superior performance in the professional exercise.

Constructivist approach

This approach has been developed in France, with the objective of overcoming some lacks of the functionalist approach. It is not based on the functionalist analysis, but in the study of difficulties of the laboral-professional processes. Based on this study, the curriculum of competencies is built, so that people learn to face such difficulties, in a successful way. Curriculum makes people to form competencies according to the professional and organizational dynamics, but having in mind their own expectancies and objectives (something not considered in the functional approach).

In the current evolution of this approach, competencies are considered as processes that go developing along with the studies plan. This is positive because they are tackled in an integral form. However, it is sometimes difficult to settle on learning in each formative space. This is because the responsibility about formation is not clear among teachers. If this approach is worked, there is a need of full collaborative work among teachers, as well as frequent meetings of coordination. It is also precise a continuous follow up to students´ learning, in each formative space and in the transition from a formative space to other one.

Socioformative approach

This approach has been developed from the contributions of several Latin American and European researchers. Its emphasis is set on innovating the educative processes so that they are at the same level of the great challenges of humanity. It seeks that integral formation and the development of competencies get generated from the ethical project of life. It differentiates from the other approaches because it gives great importance as to institutional philosophy in the determination of competencies to be formed, as to analysis of problems (challenges) that students must learn to face in the present and in the future.

The approach looks for the change of traditional, educative practices through the deconstruction of the rooted mental models in people, which block innovation. The other approaches, on the other hand, come out from the requirements of the context and they expect the curriculum to be an answer to the established requirements. The socioformative approach has also in mind the demands and the requirements of the context, but from an oriented what for, based on creating and innovating to transform the context.

In the practice, there are not pure approaches: In the curricular management it is common to find contributions of several approaches which, if it is done in a conceptual and methodological rigor, it can be a positive factor to face the several requirements of a determined educative institution. However, what many times happens is that there is an unarticulated and woven eclecticism of several contributions. This generates curricular designs without internal or external coherence. Therefore, a curriculum with contributions of several approaches which could be complemented among them can be managed, to be linked to the institutional philosophy and that produce synergy to reach the goals of a determined program of formation. (Synergy is an articulation or complementary aspect of two or more aspects. This can produce a major impact in the achievement of a goal, in comparison with the impact that would generate each aspect separately).

Frame 2.2 comparison among the several approaches of curricular management of competencies.

	Functionalist approach	*Behavioural approach of curricular design*	*Constructivist approach*	*Socioformative approach*
Goals	People carry on productive functions through units of competence.	Development and consolidation of key competencies, as essential for the person as for the success of social and/or entrepreneurial organizations .	Development and guarantee of competencies from the frame of processes	· It makes emphasis to form competencies from the ethical project of life. It considers the abilities of complex thinking. · It promotes the development of the entrepreneurial spirit, by considering the social, historical, political, recreative, disciplinary and scientific context
Structure	Modules in which students are expected to learn by doing tasks and similar activities to those of the laboral, professional and social context.	Very well specified courses, with a high emphasis in the technology of learning and the development of competencies from a more organizational vision. Materials of a programmed teaching.	Modules, integrative subjects, and projects.	· Formative Projects, in which it is intended the students to learn and reinforce the competencies of a complex mind, the work in the ethical project of life and the resolution of problems of the context from.
Emphasis of the methodology of curricular design.	To identify main and secondary functions which demand the laboral, social and scientific processes through the functional analysis to determine and norm the competencies and to build modules.	To identify the competencies by watching to highly competent professionals, to register how they develop.	To determine difficulties in processes of the organizative dynamics.	To depart form the study of the formative processes of the institution and to search the innovation from the sistemical analysis, by understanding the challenges of the context and developing abilities of complex thinking in all the educative community. To build dynamic processes in the surroundings and to determine the competencies from these ones. To use the educative action research as a dynamics of curricular design by competencies and addressing from the fifth discipline.

Which are the traditional phases of curricular design by competencies?

The methodologies of curricular design by competencies are multiple and diverse. It is due to the fact that there is a wide variety of purposes, epistemologies and pedagogical models that are for reference. However, behind all this ample variety of proposals to design the curriculum, there is a tendency to accept that, in general, there are at least three essential phases in this process , whose central components tend to be found in the diverse methodologies and approaches. These phases are: contextualization-profile, plan of formation and implementation, (fig 2.3) (see, for example, Maldonado, 2001, 2006 and Tobón, 2002, 2009a).

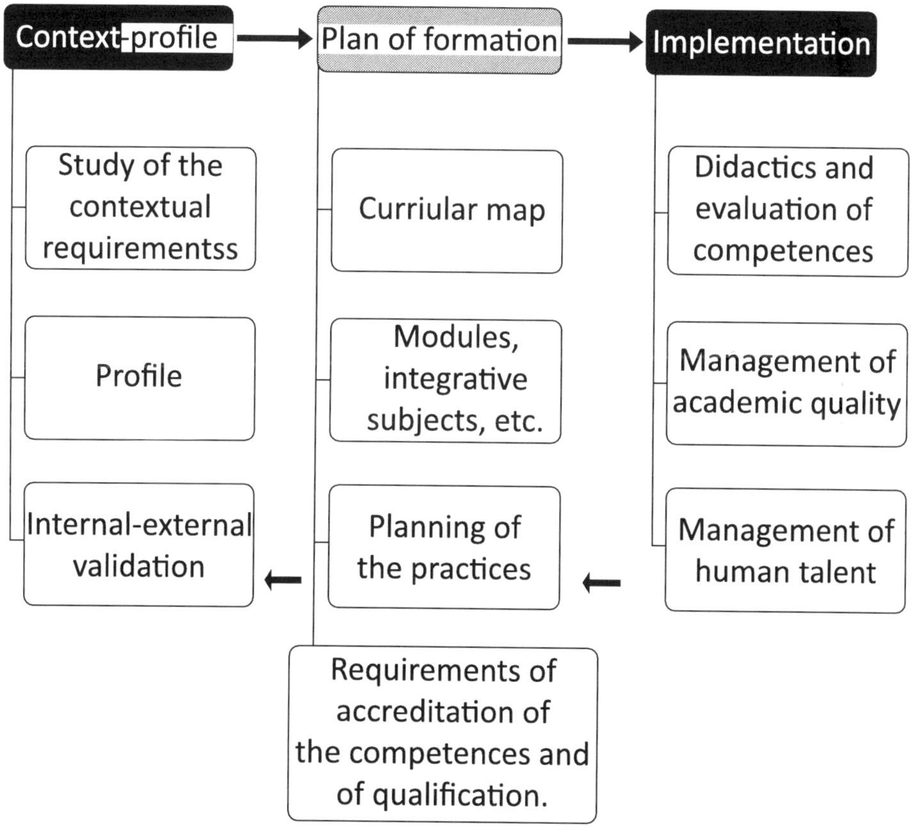

fig. 2.3 Most typical phases of curricular management by competencies.

Phase 1. Contextualization-profile

It is widely accepted that every process of curricular management by competencies begins with two central activities: the study of the context and the building of the graduation profile; according with the characteristics of the program which is intended to be established and the cycle in which it is centered. The study of the context consists on determining the requirements of the legal, institutional, laboral-professional, disciplinary-researching and social context, This is done, to establish a profile of graduation of the program, to build the plan of studies and to implement the learning and evaluative activities.
The profile is a key component in every educative program by competencies, because in it the basic, generic and specific competencies expected to be developed to strengthen the students, are described. The profile is expected to be elaborated according to the challenges of the institutional and external contexts, and after, it is socialized with the educative community to ensure its pertinence.

Phase 2. Plan of formation

Once the study of the context has been done, and the graduation profile has been built and validated, there is a second phase of design in an educative program by competencies. This phase consists on elaborating the map or curricular net. It means, the general plan that is going to orient to the students about how to develop and strengthen the established competencies in the graduation profile, from the beginning until the end of the program. To achieve this, the curricular map describe which formative projects the students are going to course, in which periods, how long every period lasts, and how many credits every formative project has. Additionally to this, the net describes the total credits for each academic period (it can be a two-month, three month, biannually or annually period). And for all the program, together with the professional practices that are carried out, their credits and and their respective requirements for titling, among other aspects. This is also the base to administer the program and to carry out the teaching exercise.

Another important component of this phase consists on planning each one of the formative spaces proposed in the curricular map. Generally, there is work by competencies with modules. However, in the Socioformative approach it is done with formative projects. This requires that professors plan each one of the formative projects in their fundamental elements before beginning each period of formation. A formative project, as it is explained in other moments of this book, consists, on one side, in the planning of a project, being meaningful for the students, with the objective of forming at least a competence of the graduation profile, with an important number of credits (it is recommended to have at least four credits by every formative project), and, on the other side, the team work and in a net among teachers of the program, looking for the articulation of theory-practice.

Phase 3. Implementation

As its name indicates, this phase consists on implementing the formation plan, established in the previous phase with the students. Therefore, it is necessary to count with human, professor, pertinent talent and the different required resources for the formation spaces. It must be considered that the teachers have to be trained and formed very well in all the competencies approach, and that they have to develop with suitability to apply didactic strategies and of evaluation, according to this approach. Likewise, in every space of formation (which can be modules or formative projects) the teachers must plan concrete, learning and evaluative activities with the students, with their respective materials of study, including laboratory equipment, probationary, etc. The institutions are recommended to do a continuous follow up to professors during the implementation phase. This is with the objective of helping them in the process and to solve difficulties that might appear, which are frequent in every innovation.

MANAGEMENT OF THE FORMATION BY COMPETENCIES FROM THE SOCIOFORMATIVE APPROACH

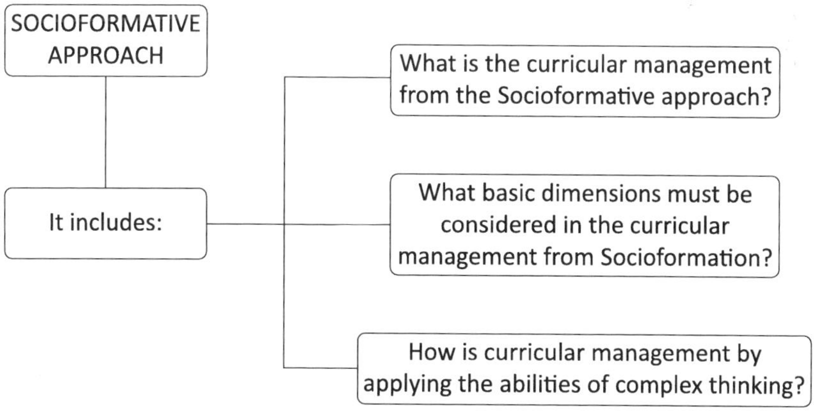

What is the curricular management from the socioformative approach?

The socioformative approach conceives curricular management as the achievement of formative projects with the students to mediate and to ensure the integral formation and the development of the expected competencies according to the challenges of the context, through committed and competent principals and teachers, which can be shown in the consolidation of social tissue, the socio-economical development, the equilibrium and the sustainable environment, and the scientific advance.

In this order of ideas, the curriculum must be a process of participative construction and, with leadership, self-organizative,dialogic, recursive and hologramatic leadership, able to search integral, human beings, with a clear ethical project of life, with entrepreneurial-researching spirit, and with suitability to face the life and the social challenges (fig.2.4).

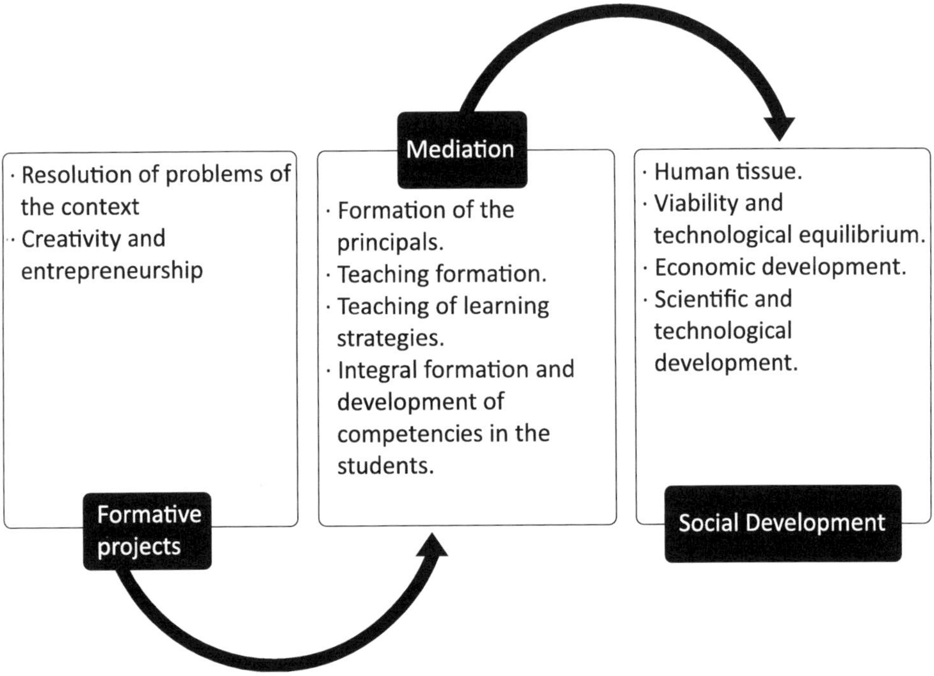

Figure 2.4. General process of curricular management from integral formation and the complex thinking.

From the socioformative approach, the curricular management in education differs from several perspectives and ways of setting out the curricular design, which is frequently followed by some educative institutions in minor or in great measure. Next, each one of these perspectives and its differences with the socioformative approach are explained (fig 2.5).

1.- Curricular design from the requirements of the context on. In the educative institutions it is frequent to find processes of curricular design based almost exclusively in the study of the external requirements. This makes the design to be basically an answer to what the legal, political, social, professional and researching contexts demand. This produces a lose of identity in the same institution, due to the fact that the educative proposal is basically an answer to what the context requires.

The socioformative design takes into account the requirements of the context, but it tackles them since the construction of a very clear institutional philosophy, with proposals of innovative formation, to which it articulates the requirements and demands of the surroundings. In this form, it can be said that the socioformative approach looks for a proposing curriculum, which may consider not only the contextual requirements, but also to look for overcoming them and to formulate new proposals; even to realities which do not exist yet, but are considered essential since the institutional philosophy, by respecting the universal values.

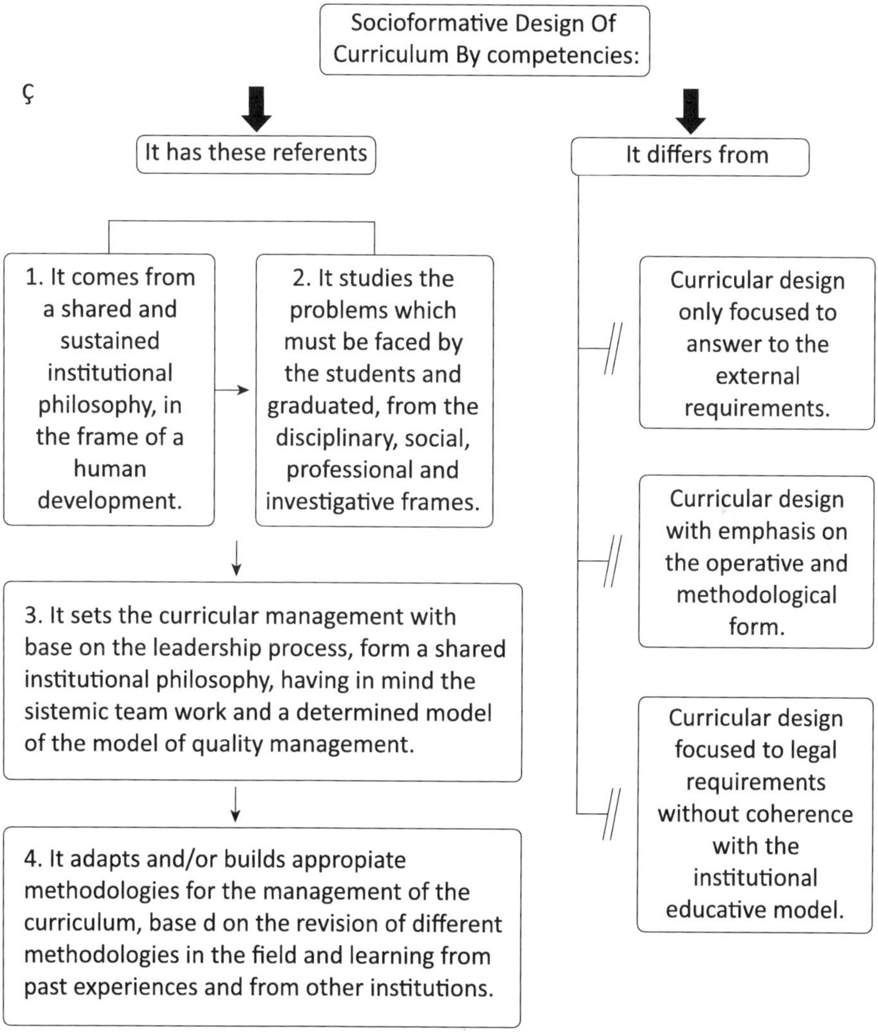

Figure 2.5 Differences of socioformative curricular management with some common perspectives of curricular design.

2.- Curricular design with emphasis in the operative form. There are sometimes some institutions that, when they design an educative program by competencies, get centered to follow a determined methodology already established in this area. From that, they build the profile, the curricular map, the formative projects, and the teaching-learning-evaluation process is implemented, without any reflection about the pedagogical model to be followed or the orientation of values and institutional principles. This frequently leads to have a curricular design which corresponds very well to a methodology of design, but without pertinence towards the institutional philosophy.

For instance, in Argentina, CIFE Institute reviewed a curricular program established by cycles in the engineering area. It found that all the program was built by always following the functionalist methodology. A functional analysis with experts about those functions which had to be performed by engineers, was done as well as by technicians and technologists in that area. Then, competencies based on the identified functions of that area were built. With all this information, the profiles of every pre requisite cycle were done, and the modules were built, by looking forward to have each one of them to answer to a determine unit of competence, which was based on required functions of the context.

This program of curricular design by cycles and by competencies was then compared with the institutional philosophy. The result was that there was not coherence nor pertinence, due to the fact that the institutional philosophy privileged several generic competencies which did not appear at any part of the profile, because they were not prioritized by the experts along the functional analysis. It was the case of innovation, the competence of multiculturalism and the competence of biophysics (to have a healthy life through sport and recreation). Also, even when the institutional philosophy was to form entrepreneurial professionals with a futurist vision, this did not remain established in the profiles, nor in the modules of every prerequisite cycle. Therefore, the emphasis was set to answer to the required functions by the productive sector in which the professionals of the engineering had to work. The problem in this area was not the use of the functionalist analysis, but the fact of not having considered the institutional philosophy. This is a frequent problem when programs by competencies are designed.

3.- Curricular Design focused on legal requirements without coherence with an institutional educative model. This is another common form of setting out the curricular design by competencies. It consists on carrying out the curricular design looking forward it to meet the minimum requirements and current norms entailed by law in education; without following in a coherent and appropriate form a determined educative model. We find, then, institutions where processes of curricular design have been made, which meet very well the legal requirements, but without the support of an educative model. This means: the curricular design does not answer to any type of person to form, to a determined conception of culture, of human society, of education or learning environments. Neither does it articulate the vision, mission or institutional principles which must orient the design.

4.- Curricular design centered in the institutional philosophy or in the vision of the authors only. This design is opposite to the previous perspective. It consists on designing the curriculum by competencies having in mind the institutional philosophy and the built educative model, without considering in a rigorous form the current and future requirements of the social, professional, disciplinary or investigative contexts.

In this line are also those designs of programs which are carried out from the expectancies, opinions, visions, and academic considerations of one or several professors, without considering the systematic study of the context. In these two forms of proceeding, the curriculum is an internal design, coherent with the institutional philosophy or with the visions and academic stances of those who establish it, but with a huge lack of pertinence towards the requirements, problems or challenges of the society.

5.- Emphasis to design a curriculum but not in its management. Traditionally, the institutions have focused on how to design the curriculum, and this makes that once it has been achieved, the topic is never referred again. In the socioformative approach, it does not only design the curriculum. It also looks for creating mechanisms to continue managing it, so that it can be implemented with quality, and it to continue being continuously improved with a coordinating team of the process in the institution, faculty and program.

The socioformative approach seeks to tackle curricular management in a different form of how it has been done in education: basically, an activity of some professors and from their own experiences, academic formation and philosophy. The socioformative curricular management has a special characteristic because it looks for understanding the meaning of the curricular process regarding the formation of integral and honest people very well, from processes of teamwork with synergy.

This lasts consists on looking for common goals through the complementation of attitudes, values, capacities, abilities, skills, and knowledge among the people in a team. When this impact is achieved, with the searched goals, then, it can be said that the team has synergy. Also, the curricular management seeks the process to be pertinent in a global, ecological plan, in relation to self-fulfillment of each person and his ethical project of life; and, at the same time, to consider the construction of social tissue, the professional suitability, and the contribution to the big problems of humanity (Morin, 2002 a; Tobón, 2009a), the entrepreneurial of social and corporate projects, and the search of sustainable development and ecological equilibrium.

The curriculum tends to be tackled in a deterministic manner, organized in a lineal form, and under a low grade of dynamics, which makes difficult its flexibility and maintenance of appropriateness of formation. The same traditional science has privileged this paradigm in the building of scientific theories, leaving aside uncertainty, the diversity, the mistake, the change. The complex thinking does not oppose to systematic, strategic, methodic curriculum or it is to tackle processes of order towards formation. On the contrary, it looks for taking into account all these aspects, but complemented with a complex vision of man, of the world and of humanity. This vision consists on interwaving the formative processes, to assume them in a continuous change and to re-organize them so that they do not lose pertinence, to be closer from different educative perspectives, articulated among them, and trying not to fall down in reductionism which appears when there is only one kind of work, with a single frame of reference.

The socioformative curriculum is not a curriculum of chaos, or of uncertainty in education. It is, according to professor Morin (1995), conscious, intentional and sustained planning and management , supported by how people are going to be formed, in a determined historic, social, political and economic moment. It goes through phases with a certain logic sequence (in other cases, they are given in a parallel form, as the building of formative projects with the construction of curricular map), by having in mind the tackling of processes of uncertainty and its strategic confrontation. Some situations of chaos are due to legal changes, unexpected changes in the requirements of the context or to emotional changes in people.

In the same way, to think the curriculum in a complex form is also to join the study of each one of the stages of curricular management separately, with the systemic analysis of all the curricular structure, with the objective of being able to understand the methodological general orientation and the What for of each one of the parts. This is frequently a lack in people and teams who carry out curricular designs by competencies, which can be evident in the lack of articulation of the diverse components of curriculum. For example, in a recent visit to a Chilean university to evaluate their educative project, CIFE found that the program of studies of the Business Administration Career, had been built without a methodological coherence. It could be seen because each one of its components answered to different goals, without a unified or oriented What for. A fact that illustrates this is that the profile of the program of business administration sets out to form people with competencies in English; however, this subject does not appear in the map nor is it a requirement for titling.

The complex thinking in curricular management accepts that the management can be done since a unique methodological approach and not necessarily since an inter and trans disciplinary construction. Nevertheless, even though there is an only methodological approach, it is precise to study several approaches first to select the most suitable. Since complex thinking, it is also recommended to articulate contributions from other methodological approaches to answer in a more comprehensive form to the diverse dimensions of curriculum by competencies. The challenge is to meet inter and trans disciplinary methodologies which increase the possibilities of impact in the formation, and to help to form integral and competent people.

Last, curricular management from the complex thinking assumes that curriculum is, complex. A term that comes from complexus which means tissue: an articulation and integration of parts in a more dynamic process, which is at the same time a unit in diversity (Morin, 1995,2002b). The diverse parts of curriculum are linked together, in the sense of the Pascal postulate (Dominguez, 2006; Tobón, 2005):

> *All the things are caused and cause, helped and helpers, mediate and immediate and all supported by a natural union and insensible which links the farthest ones and the most different ones. I think it is impossible to know the parts without knowing the whole or to know the whole without knowing particularly the parts.*

This means that to understand the curriculum, one must understand each one of its parts. And to understand each part, one needs to relate the parts among them and to understand the curriculum as a whole.

For instance, the graduation profile only has one sense as long as there is understanding of its own nature and it is related with the other parts of curriculum such as the educative institutional model (which establishes what type of human being to form and for which society), the curricular map (or curricular net), the formative projects, the didactics, the learning sceneries, evaluation and certification of competencies. That is why to design the curriculum, it is required first to have a systematic and well planned methodology of all the process, in which it can be included the design of each one of its parts in an interrelated form and to think at every moment in a what for. It can be done in the same form which happens when a dress is made: each part is woven by having in mind all the set. The vision is set to have as a final product a dress, which satisfies the needs and expectations of each person.

Which are the basic dimensions which must be considered in the curricular management from the socioformation approach?

According with what was analyzed in the previous paragraph, the curriculum by competencies since complex thinking has the following dimensions:

1. It is a formative project.

2. It searches to form integral human beings with a solid project of life
3. It intends to develop the entrepreneurial spirit
4. It has as a challenge to form generic and specific competencies necessary for the human and social development.

The curriculum is a formative project

In the socioformative approach , the curriculum is always a project, because it is continuously projecting in present towards the future. It never has an end and it is established as any project, with planning, execution and communication of the goals.

Additionally, from the socioformative approach, the curriculum is oriented to form. It means: it searches to generate meaningful learning and evaluating experiences so that people give form to their different potentials since the multiple intelligences and that they strengthen them continuously, besides the learning of academic knowledge.

Curriculum aims to the ethical project of life

It refers to the process in which each person plans, lives and values his life with base on goals which respond to his vital needs of growth and development; according to construction, reflexion, appropriateness, and application of universal values, by considering his being, the social tissue, the economical development, the ambiance and the cosmic mother country in its whole.

In the socioformative approach, it is searched that people construct and guarantee their ethic project of life, and from this, they carry out their general and specific formation. In this manner, the learning, teaching and evaluative activities are oriented around the vital needs, the personal goals of development, the values and ethics, which avoids a high grade of fragmentation which tends to appear in the formation by competencies, due to the lack of an integrative and articulating component. In the socioformative approach, the goal then is not to form competencies, but to form people who manage their personal self-fulfillment with a social, environmental, professional and organizative commitment, and in this frame, they set into action their competencies.

The curriculum searches to form the entrepreneurial spirit

The entrepreneurial spirit refers to construct and to implement projects to satisfy determined needs or to solve problems in the personal, familiar, social, laboral and organizative entrepreneurial order with perseverance and a challenging sense until reaching the expected purposes. To learn how to undertake is to develop and to strengthen attitudes, values, knowledge, abilities, skills to solve problems, effectiveness and efficiency.

The curriculum, in the complex thinking, searches that people develop and support the entrepreneurial spirit in a progressive form, so that they become principal actors of their own life and reach their whole fulfillment by participating in the solution of problems and creating new realities in the social, political, economic, professional and organizational-entrepreneurial area. Thus, the curriculum makes the strategies to be explicit to reach such proposal not only in the profile, but also in the curricular map, in each one of the cycles and in the formative projects. In this manner, there is also an answer to one of the most important educative policies in international scale, as it is formation to self-employment, which is included in the know how to undertake.

Curriculum settles on generic and specific competencies

From the socioformative approach, the competencies were analyzed in chapter 1. They refer, in a concrete form, to integral performance that people develop and set into action to analyze, to understand, and solve problems of different contexts, with suitability, creativity, flexibility, metacognition and ethics by confronting uncertainty strategically. They are the form of how the different dimensions and components of curricular management are concreted, because from competencies, the profile, the curricular map the formative projects, the didactic strategies, evaluation, and accreditation of reached learning in a determined cycle are configured.

How is curricular management when the abilities of complex thinking are applied?

From the socioformative approach , it is very important that curricular management is carried out in all its processes and axis with base on the essential abilities of complex thinking. These abilities are: self-organization, metacognition, holo grammar, recursion and dialogic ability.

Self organization

The self organization is a process by which a system searches and builds its differentiation (identity) in relation to determined purposes (goals), (autonomy) through relations of dependence in continuous change with the context (dynamics of dependence). The curriculum, thought from self organization, is a project of formation in a continuous improvement to generate impact in the integral, human formation and in the competencies (goals), since institutional philosophy of human formation (dynamics of autonomy), looking for the identity of the educative program of the institution (identity), with base on systemic relations continuously modifiable, with the challenges and requirements of the institutional, social, professional, and investigative context (dynamics of dependence).

In this sense, the challenge from complex thinking is to assume curriculum as a human process which must continuously self organize, in such a form to renovate the profile, the curricular map, the formative projects, and the policies of formation-evaluation from the philosophical principles of the institutional educative model(internal dynamics of autonomy), through changing relationship with the requirements of the different contexts (dynamic of dependence).

Metacognition

The Metacognition consists on achieving a goal by reflecting upon the own performance and self-regulating such performance, in such a way that the purpose is achieved with the major quality as possible. In the curricular management, this means that the curriculum must be done with the continuous practice of metacognition, with the objective that people make consciousness of what they do, they find sense to the management, and self regulate their performance by looking for the best quality in a perspective of continuous improvement.

Holo grammar

The holo grammar principle of complex thinking states that, to understand the alive phenomena, it is essential to analyze how the whole is present in each one of the parts (Morin, 1995).

For example, the human beings form a society, but the structure of society is present in each person. Additionally, we have a body formed by cells, and in each cell there is the structure of all the body in the genes.

With respect to curricular management, by competencies, this means:

- To apply the formative projects (parts) by having in mind the institutional educative model (the whole).
- To look for the students to reach the graduation profile (parts) by taking as a base the institutional educative model (the whole).

Recursive

The cause performs over the effect, and the effect performs over the cause, by breaking the lineal causality proper of traditional, positivist epistemology. The processes self regulate with base on the information and the context. In relation to curricular management, there must be some mechanisms so that curriculums self feedback continuously form the evaluations of the teachers and students and that this serves to be modified and changed, in search of a major flexibility and pertinence, without a major rigidity and de contextualization. If a pertinent formation is given, there will be a formation of quality; and this will make curriculum to be valued and consolidated.

Dialogic

The antagonic principles must be joined by searching their complementarity to understand complex phenomena and to generate change and innovation. In the curricular management exist several pairs of antagonic principles, among others: direction and flexibility, professional orientation and scientific orientation.

The dialogic principle invites us to search complementarity of these principles as a base to a more complete boarding of curriculum. Therefore, it is necessary to have a basic and common direction for everybody, with a logic sequence of a certain number of formative projects so that the program becomes suitable. But it is also precise to have flexibility: It means, the students are able to configure their studies according to their expectancies, straightforwardness of time and goals.

Likely, it is required to join and to search the complementarity between professional formation and scientific formation, which is achieved when investigative formation is promoted in a transversal manner in the curriculum, it is worked based on problems and projects and it is emphasized in the creativity and innovation.

ACHIEVED LEARNING IN THE APPLICATION OF THE SOCIOFORMATIVE APPROACH

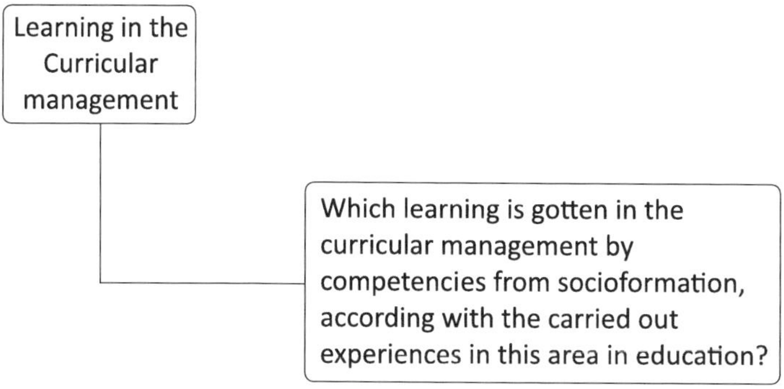

It is almost 15 years of application of the socioformative approach in the educative institutions, and the learning is multiple and varied. Next, these learnings are synthetized so that they are taken into account as by educative institutions which have followed the socioformative approach, as by those others which have followed other approaches or which are about to begin the process.

Learning 1. Curricular management must be a simple process

The institutions have frequently developed very detailed and sophisticated models about curricular management, which have not been able to be applied later on, or which have required of a very long time to be carried out to practice. This usually leads to paralysis, because it complicates to continue with a model of many activities when there is also the task of formation, evaluation, research and to do extension. That is why, the socioformative approach has searched with several investigations to do all the process more simple, by eliminating multiple factors which tend to complicate more than to produce quality in the learning, teaching and valuation. For example, the study of the context, curricular map, and the formative projects have been simplified, to leave only the essential one.

Learning 2. The Curricular management must tackle the different components of the system

Very frequently, the processes of curricular management in the educative processes are very detailed in their activities, but they forget components that later on are essential in the process which withdraws the impact to the transformation of the plans of study. For example, at the moment of planning the curricular management, in the institutions, with frequency, they forget the follow up, the evaluation, of quality management, of formation about principals, and the process of certification of the competencies in the students. In the socioformative approach, these aspects are considered essential in the management of the curriculum, because the experience has taught us that they are responsible of a great impact in all the process.

Learning 3. A key factor is that there must be leadership of the process

Experience has shown us that the goal uses to consist on having plans of study by competencies and when it is achieved, the process is forgotten.

This is due to the lack of a clear leadership, in which there must be people of an educative institution which is in charged of that, and to have it assigned in their functions and responsibilities. In the socioformative approach it is emphasized, according with this learning, to have a team to coordinate all the process, with clear responsibilities, as well as in the institutional plan, as in each faculty and academic program of pre grade and of post grade. These teams do not have to work out solely in the process of curricular design, but also in the phase of implementation, during the follow up and the evaluation, and to have continuity to ensure the continuous improvement.

Learning 4. Curricular management is important to be articulated with the educative model

It has many times been intended to implement processes of curricular management by competencies without making changes in the educative model of the institution. It is sometimes implicitly thought that, if something is done, the institutional philosophy is changed or that competencies can be worked since any model without need of making changes. For experience, we know that, to have an impact of the competency approach, this must be integrated as such in the educative model, by obviously respecting the educative model and the philosophy of the institution as such. From this, in the socioformative approach, an aspect of management is to look for the competency-based formation to be incorporated as such to the educative model of the institution.

Learning 5. There must be strategies of change in the institution

Change is not given by the simple transformation of the plans of study, neither is it done by the training teachers. The change is given when the teachers are commited with the process and assume it, by implementing processes of change in their teaching as it is the case of the educative, action research. The training processes should be more processes of research to transform the teaching practices, having as a referent the complex thinking and the competencies. If it does not happen, there will not be changes in the educative reality of the institutions.

METHODS TO BUILD UP THE CURRICULUM BY Competencies

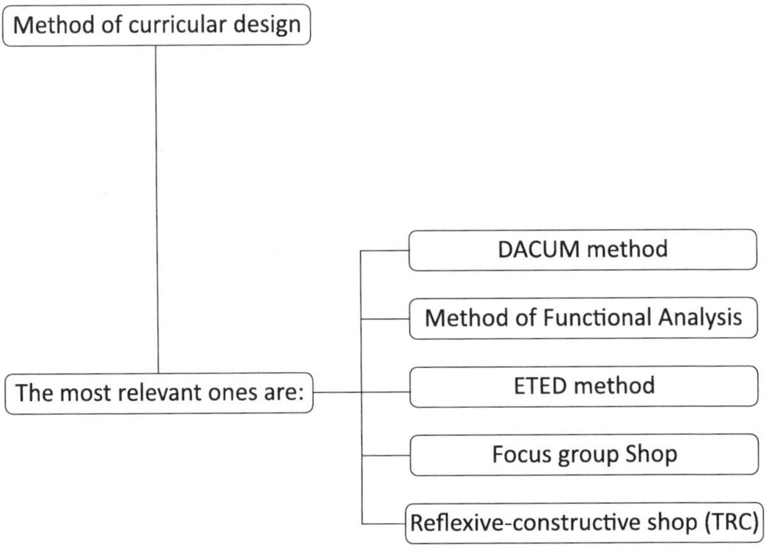

There is a wide variety of methods to construct the curriculum by competencies. However, in Latin America, the most used are the following ones: DATUM method, Method of Functional Analysis, ETED method, Focus group Shop and Reflexive-constructive shop (TRC). Among all these methods, there are some convergent points and there are some notable differences. The main point which they share is the study of the requirements of the external context to establish the curriculum by competencies. Concerning to differences, these can be appreciated on frame 2.4.

From the socioformative approach, the reflexive-constructive shop is preferred as an essential method to orient the curricular construction, with the purpose of forming and evaluating competencies, by considering the complex thinking and the metacognitive practice. In this method, the different tools of qualitative and quantitative research are considered, which facilitates all the process because people do not have to learn sophisticated procedures, as it happens with the other methods.

Next, the mentioned methods to design the curriculum by competencies are described and they are compared with the TRC.

DACUM Method

The DACUM (Developing a curriculum) is one of the most traditional methods to identify and to construct in a practical manner the competencies in a curricular process. It is done, by considering the requirements of the context through a group work shop, with people highly expert of the activities and demands which people must face in the labor-professional environment. (Norton, 1995, 1997).

The principal advantage of DACUM method, above other methods (such as functional analysis) is that it makes possible general guidelines around construction of curriculum (Mertens, 1997) and this implies that from the beginning competencies are determined by thinking in the program of formation, which does not happen with functional analysis, because in it, the occupational functions are analyzed on one side, and then, the modules and plans are designed, which frequently implies new adaptations and modifications. In DACUM, on the other hand, the process is direct and does not imply different moments.

Traditionally, the DACUM has been used with the following steps:

1. Selection of the laboral-professional area. To achieve this, it is required to build a curriculum of formation by competencies.
2. Invitation to 8-12 expert people of the selected laboral professional area.
3. Group reunions with the invited people through two expert facilitators in the DACUM methodology.
4. The experts explain the DACUM methodology to invited people.
5. The experts use group dynamics to obtain the information of the attendants with respect to the area of interest.
6. The laboral-professional area is described with the contributions of the attendants until having an agreed common base.
7. The general activities are identified (or general functions) and concrete tasks through colorful tokens which are displayed on a wall, board or blackboard. As soon as the participants go collaborating the tokens are reorganized or they go changing some for others.
8. The units and elements of competence are described and they are organized in order, with the tokens.
9. The related knowledge, the behavior, attitudes, teams, tools, materials and elements of safety are identified.

Mertens (1997) describes the following series of problems which appear in the DACUM traditional methodology.

a. It does not part from the global function of the company (or of the profession). This generates the risk of not including more transversal competencies, such as the capacity of the organization to work by objectives.
b. it is emphasized on decomposing functions in tasks, and this can lead to fragment the occupational or professional performance.
c. Little importance is given to malfunction, to problems, to

unexpected situations and to non-required operations.

d. There is too much trust in contributions of experts. It is necessary to consider other valid sources.

e. The task is described in terms of actions and results. Attitudes, leadership, communication and other elements related with the capacity of social interaction do not appear like that, which for organizations are essential factors.

The TRC retakes from DACUM its emphasis in the participative shop, its dynamics of work through file cards or observable cards and the need of organizing the competencies in an interrelated form. However, TRC contributes to improve DACUM, when it promotes more reflection about educative practices and to generate changes in the mental, negative models through the metacognitive practice and action research: it searches the design of a whole educative program and not only the profile and the curricular map, and it focuses from management to academic quality.

Method of Functional Analysis

It is a method focused in determining the competencies which people must posses in an occupation, through the analysis of the required functions to fulfill the key purpose of a determined area, organization or set of organizations. Different from DACUM method, which centers in the tasks of an occupation, the functional analysis studies the relationship among functions and the organizational context, taken as a system, then, the relation parts-whole and whole-parts. This method has been promoted by the National Vocational Qualifications (NVQ) of the United Kingdom.

Functions express what a professional must be able to do under normal conditions of work, with the objective of fulfilling with a key goal. In general terms, those ones indicate the activities that people must perform in a determined laboral area. In more concrete terms, the functional analysis has as a basis (Zúñiga, 2003): a) the key purpose of an area. B). the functions of first level that workers and professionals must do to reach such a purpose (units of competence) and, c) the functions of second level (elements of competence or professional exercise). With respect to the methodology of elaboration, it is suggested to review Irigoin and Vargas (2002).

The basic steps of functional analysis are:

1. To establish a group of experts with workers (or professionals), people who know very well the occupation and people who investigate in the area.
2. To plan and agree an action plan, with goals, activities, and timetable of work.

3. To determine and agree with the groups of experts the key purpose of the area, department, organization or set of organizations where the analysis is going to be applied. From this key purpose, the units of competence and the elements of competence are determined.
4. To build up the functional map, by identifying the units of competencies as follows: which competencies do people require (not the teams) to reach the key purpose?
5. Once reviewed and agreed the units of competence, to identify the elements of competence through the question: which activities must people do (not a group), so that the unit of competence is achieved and it to contribute to the key purpose?
6. To review and to agree the elements of competence.
7. To write down the criteria of performance in each established element of competence, by considering the occupational-professional context. These criteria allow to evaluate the quality in which each element of competence is executed.
8. To describe the rank or field of application (context or contexts in which the elements of competence is applied.)
9. To write down the evidences of performance, of product and of knowledge
10. To establish the essential, required knowledge to execute the element of competencies and to fulfil with the criteria.
11. 11. To review the units with their elements and components, and to improve and to agree about the process. In this stage, it is accustomed to consult to other external experts to ensure the pertinence and quality of the functional analysis done.
12. 12. To document the functional analysis and to socialize it.

In the frame 2.3 an example about the functional analysis, made in Colombia about the educative sector is described. This shows all the functions that are precise to carry out, with the objective of fulfilling with the key purpose of forming people in an integral form, who participate in the transformation of the Colombian Society. This functional analysis was elaborated in 2002, by the Table of the Educative Sector, which is led by the National Ministry of Education and technically coordinated by the SENA (National Service of Learning). The analysis is done with base in the concerted participation of professionals of education, representatives of professional associations, institutions of basic and middle education, technical institutions, universities, boxes of familiar compensation and unions of production.

Frame 2.3

Key purpose: To form people who participate in an organized way in the transformation of culture and Colombian society, in a frame of justice, equity and environmental sustainability.		
Key function A: To generate knowledge with base on an integral vision of the people´s needs and of national development.		
Function level 2	*Function level 3*	*Function level 4*
A.1. To form investigative communities with base on complexity of problems to be solved.	*A.1.1. To integrate nuclei of investigation with base on the perspective of the disciplinarian, occupational or professional area.*	*A.1.1.1 To identify the area or areas of research in the institution with base on its plans of action and the context.* *A.1.1.2 To establish plans of action in research, In accordance with the structure, the human talent, and the resources of the institution.*
	A.1.2 To structure systems of information and communication of the research groups, in coherence with the requirements of the programs of science and technology.	*A.1.2.1 To ensure the availability of infrastructure of information and communication with base in the interaction among groups or research.* *A.1.2.2 To organize the validation and diffusion of products and services of the researching communities with base on national and international protocols.*
	A.1.3 To organize services of extension with base on the requirements of productivity and quality of enterprises and organizations.	*A.1.3.1 To design processes of technological improvement, with base in the detected problems.* *A1.3.2. To establish the service of consultancy with base on the requirements of the enterprises and social organizations.* *A.1.3.3 To structure systems of indicators of innovation and technological development in coherence with the state of art in the selected areas.*

	A.2.3 To structure programs of formation with base on the established competencies and in the institutional educative policy.	*A.2.3.1To set out the curricular structure with base on the logic of area object of formation, in the needs of population and in the institutional, educative policy.* *A.2.3.2 To organize plans of study with base on an approach of knowledge construction and of knowledge dialogue.*
	A.2.4. To plan the formative processes according to the institutional parameters.	*A.2.4.1. To organize the teaching-learning-evaluation strategies according to the planned learning results.* *A.2.4.2. To design learning environments with base on the proposed results and in the characteristics and requirements of the students.*
Function Key B: To generate the educative service with base on the purposes and strategies of the plan of a definite development.		
Function level 2	*Function level 3*	*Function level 4*
B.1 To plan the educative service, in coherence with a human development approach.	*B.1.1.To formulate the educative project with base on the institutional philosophy, the educative policy of the state, the current norms and the needs of the surroundings.*	*B.1.1.1.to define the teleological of the educative project with base in the results of the diagnostic.* *B.1.1.2 to draw up the institutional strategy in coherence with the established priorities in the educative project and the results of reference.* *B.1.1.3. To orient the formulation of parameters and indicators for the evaluation of the educative project, with base on the institutional philosopy.*
	B.1.2. To determine the parameters and signs for the evaluation of quality of the educative service.	*B.1.2.1. To fix the technical criteria and the rates of assignation of personnel to give an educative service of the state, by having in mind the particular aspects of each region.* *B.1.2.2. To establish the requirements for accreditation and or certification of institutions, territorial entities, programs and people according to current norms.* *B,1,2.3 to establish rewards for the districts, municipalities, and*

Key purpose: To form people who participate in an organized way in the transformation of culture and Colombian society, in a frame of justice, equity and environmental sustainability.		
Key function A: To generate knowledge with base on an integral vision of the people´s needs and of national development.		
Function level 2	*Function level 3*	*Function level 4*
A.1. To form investigative communities with base on complexity of problems to be solved.	*A.1.1. To integrate nuclei of investigation with base on the perspective of the disciplinarian, occupational or professional area.*	*A.1.1.1 To identify the area or areas of research in the institution with base on its plans of action and the context.* *A.1.1.2 To establish plans of action in research, In accordance with the structure, the human talent, and the resources of the institution.*
	A.1.2 To structure systems of information and communication of the research groups, in coherence with the requirements of the programs of science and technology.	*A.1.2.1 To ensure the availability of infrastructure of information and communication with base in the interaction among groups or research.* *A.1.2.2 To organize the validation and diffusion of products and services of the researching communities with base on national and international protocols.*
	A.1.3 To organize services of extension with base on the requirements of productivity and quality of enterprises and organizations.	*A.1.3.1 To design processes of technological improvement, with base in the detected problems.* *A1.3.2. To establish the service of consultancy with base on the requirements of the enterprises and social organizations.* *A.1.3.3 To structure systems of indicators of innovation and technological development in coherence with the state of art in the selected areas.*

	B.3.2. To establish alliances with different statements according to the institutional strategy.	*B.3.1.1. To identify the indicators of management to fulfil according to the institutional goals, the educative policies and the legal norms.* *B.3.1.2. To assure the fulfilment of the indicators of management with base on institutional projects.*
	B.3.3. To manage the execution of action plans and of improvement with base on the educative policy and in the institutional laws.	*B.3.3.1. To program the activities, the resources and the required human talent with base on the requirements of operative plans and the established priorities.* *B.3.3.2 To advise to the personnel in charged with base on the established projects and the institutional, educative policy.*
B.4. To manage the necessary financial and physical resources for the development of the educative service.	*B.4.1 To manage the achievement of resources with base on the requirements of the educative project.*	*B.4.1.1. To manage projects of investment according to the institutional policy and the current legal norms.* *B.4.1.2. To organize requests for bids according to the institutional legal norms.* *B.4.1.4. To authorize the purchase of resources according to the cash flow, the assigned budget and the level of responsibility.*
	B.4.2. To administer the budget, according to the institutional policy and the current legal norms.	*B.4.2.1 To make the budget plans, according to the current legal norms.* *B.4.2.2.To control the financial execution according to the legal criteria and the institutional policies.* *B.4.2.3. To give the regulations about costs, prices, pensions, and academic rights according to the current legal norms.* *B.4.2.4. to distribute and to assign resources, according to the legal, current criteria according to the profile of the formulated projects.* *B.4.2.5 to order the expenses according to the assigned budget to the project.*

Key purpose:

To form people who organized participate in the transformation of Colombian culture and in the society in a frame of justice, equity and environmental sustainability.

Key function A: To manage the educative service with base in the purposes and strategies of definite plan.

Function level 2	*Function level 3*	*Function level 4*
	B.4.3. To keep working the required local facilities, machines, equipment and tools to the development of the educative projects.	*B.4.3.1. To contract the building or adaptation of educative ambiences for the services.* *B.4.3.2 To manage maintenance programs of the infrastructure and the institutional logistics based on the recommendations of management.* *B.4.3.3. to do the maintenance of the machines, equipment and tools which are under their inspection, according to their plan.*
	B.4.4. To administer the inventory of machines, equipment, tools, and materials according to the institutional legal norms.	*B.4.4.1. To use the resources in the assigned spaces and according to the current legal norms.* *B.4.4.2. To storage the resources, properties, and basic tools based on technical, national or international norms.*
B.5. To create the conditions for the development of the human talent which develops the educative service, according to the policy and the strategies of organization.	*B.5.1. To administer the human talent on their care, according to the current legal norms, the assigned competencies and the indicators of management.*	*B.5.1.1. To manage the selection and hiring of human talent, according to the profiles of established competences and the current legal norms.* *B.5.1.2.To create conditions which favor the organizational environment, based on the requirements of the established projects.* *B.5.1.3. To lead the development of the competitivety of the human talent, according to the established indicators.* *B.5.1.4. To manage processes of formation and updating of the personnel , according to the policies and strategies of the institution.* *B.5.1.5. To organize the processes that favor the well being of the workers of the educative sector.*

B.6. To sell the good and services, according to the educative policy and the external demands.	*B.6.1. To organize the system of information related to goods and services and their demands.*	*B.6.1.1. To gather information about offer and demands of goods and services of the educative sector.* *B.6.1.2. To form the net of information of the goods and services of the educative sector.*
	B.6.2. To offer the services and the educative products according to the requirements of the users and the organizations.	*B.6.2.1 To define the niches of market with base on the characteristics and needs of the users.* *B.6.2.2. To formulate strategies of promotion and offers and services according to the target population.*
Function C. To develop the formative projects according to the Institutional Educative Project, The current norms, and the requirements of the educative community.		
Function level 2	***Function level 3***	***Function level 4***
C.1. To implement the processes of educative administration, given by the institution	*C.1.1. To ensure the entry and the graduation rate of the students based on the legal norms.*	*C.1.1.1. To apply the entry process of the students to the institution according to the defined policy.* *C.1.1.2. To give academic certificates according to the current norms.* *C.1.1.3. To ensure the titling of the students according to the legal norms.*
	C.1.2. to achieve the implementation of the processes of formation according to the institutional goals and the current norms.	*C.1.2.1. To execute the processes of formation according to the legal norms.* *C.1.2.2. To verify the fulfillment of the institutional goals, in the processes of formation.*

Key purpose: To form people who organized participate in the transformation of Colombian culture and in the society in a frame of justice, equity and environmental sustainability.		
Key function A: To manage the educative service with base in the purposes and strategies of definite plan.		
Function level 2	*Function level 3*	*Function level 4*
C.2.to do the educative processes according to the requirements of the Institutional, educative project.	*C.2.1.To orient the teaching, learning and evaluating processes, with base on the agreed plans.*	*C.2.1.1 To assist the students according to the learning and evaluating, detected needs.* *C.2.1.2. To do inductive processes in correspondence to the criteria of the educative institution.* *C.2.1.3. To improve the processes of formation from the problems and opportunities detected in their execution and evaluation.*
	C.2.2. To give services of extension based on the requirements of the enterprises and the organizations.	*C.2.2.1 To do diagnosis and studies based on the terms of reference proposed by the social and entrepreneurial organizations.* *C.2.2.2. To give solutions to the problems set by the social, entrepreneurial and communitarian organizations.* *C.2.2.3 To organize the processes of diffusion of the scientific and technological information according to the requirements of the target population.* *C.2.2.4. To do quality tests with base on the requirements of the social and entrepreneurial organizations.*
	C.2.3. To execute research projects based on the methodological design and in the operative plan of the project.	*C.2.3.1. to process and to systematize information according to the established, methodological design.* *C.2.3.2. To validate the results of the research project according to the established protocols.*

	C.2.4 to produce the required, educative materials for the development of the formative projects.	*C.2.4.1. To select and to prescribe the educative material with base on the requirements of the target population, and the structure of the plans of formation.* *C.2.4.2. To elaborate guidelines according to the characteristics of the context and the selected format.* *C.2.4.3. To design the educative material by graphics according to the guidelines and the selected format.* *C.2.4.4. To select and to develop the software and hardware requirements of the educative material, with base on the requirements of population and the communicative intention.* *C.2.4.5. To assist technically and pedagogically the elaboration of the educative material according to the established guidelines.* *C.2.4.6. To elaborate and to develop the communicative proposal with base on the intentionality of the educative material and the learning requirements.*
Key Function D: To ensure the quality of the educative service according to the established policies.		
Function level 2	*Function level 3*	*Function level 4*
D.1. To apply the established criteria, parameters and indicators for the evaluation of the quality of the service.	*D.1.1. To direct the process of certification or accreditation of the quality of the educative service, with base on established the guidelines and the objectives of the institution.*	*D.1.1.1. To coordinate the system of quality assurance of the educative service with base on the philosophy and the guidelines adopted by the institution.* *D.1.1.2. To assure the documents of the certification or accreditation processes, according to the established parameters chosen by the institution.* *D.1.1.3. To evaluate the assurance of quality system of the educative service, with base on the established criteria.*

Key purpose: To form people who organized participate in the transformation of Colombian culture and in the society in a frame of justice, equity and environmental sustainability.		
Key Function D: To ensure the quality of the educative service according to the established policies.		
Function level 2	*Function level 3*	*Function level 4*
D.1. To apply the established criteria, parameters and indicators for the evaluation of the quality of the service.	*D.1.1. To direct the process of certification or accreditation of the quality of the educative service, with base on established the guidelines and the objectives of the institution.*	*D.1.1.1. To coordinate the system of quality assurance of the educative service with base on the philosophy and the guidelines adopted by the institution.* *D.1.1.2. To assure the documents of the certification or accreditation processes, according to the established parameters chosen by the institution.* *D.1.1.3. To evaluate the assurance of quality system of the educative service, with base on the established criteria.*
	D.1.2. To evaluate the educative project with base on the indicators of management, the competencies and the current legal norms.	*D.1.2.1. To value the performance of the human talent according to the proposed indicators and the current norms.* *D.1.2.2. To evaluate the advance of the educative project with base on the established indicators and the institutional policy.* *D.1.2.3. To evaluate the impact of the educative project with base on the established indicators and the national educative policy.* *D.1.2.4. To report according to the current norms and the established indicators of the management* *D.1.2.5. to trace plans of improvement from the results of the institutional evaluation.*

	D.1.3 To evaluate the competencies with base on the institutional procedures and the currents norms.	*D.1.3.1. To elaborate evaluation instruments necessary for the collection and register of the evidences.* *D.1.3.2. To judge the evidences towards competencies or established achievements.*
	D.1.4. To evaluate the learning achievements of the students before the established competencies or achievements.	*D.1.4.1. To make instruments for evaluation of learning of the students.* *D.1.4.2. To acknowledge the learning of the students for their location om the process of formation or certification.*
D.2. To accredit the quality of the institutions, programs and performance of the people, according to the established norms.	*D.2.1 To value the quality of the information given with base on the requirements of the accrediting or certifying organism.*	*D.2.1.1. To manage processes of self, co and heteroevaluation, according to the Institutional, Educative Project and with some guides of the accrediting and certifying organism.* *D.2.1.2. To evaluate the performance of the institutional program or person, according to the guides of the accrediting and certifying organism.*
	D.2.2. To give credits or certifies according to the given competence and the results of the auditory and the legal, current parameters.	*D.2.2.1. To socialize the results of the processes of accreditation before the educative community according to the legal authority.* *D.2.2.2. To establish plans of educative, institutional improvement according to the results of the evaluation.*

Notes: Adapted from SENA, Functional Map of the Educative Sector in Colombia, Table of the Educative Sector, Medellin, 2012. It can be seen how each function is dis rupted into more specific functions, according to the key purpose, until arriving to the individual contributions.

Some critics to functional analysis are the following ones:

a. It emphasizes in the results that people must have, but in the process
b. Although it is considered that the competencies must address a purpose, this generally is not written down in the units of competence. This makes it not to be considered and to be frequently forgotten.
c. It tends to emphasize in the occupational and laboral area.
d. The participation of people and sectors is not considered, although inexperienced in the area, they can do relevant contributions to contextualize and give pertinence to competencies, as it is the case of students, professors and social leaders.
e. It permits to establish accurately the profile of graduation of an academic program, but it does not get the elaboration of a plan of studies, which is considered like a separate process.
f. It does not tackle the context in its evolutive processes, nor does it consider the situations of variability or incertitude.

ETED Method

The ETED method (Use-Type, studied in its dynamics; in French: for example: Emplois Types En Dynamique) has as a purpose to create profiles and educative programs aligned to requirements of laboral dynamics (Meriot 2005)., with the objective of improving the formation, and link of human talent to organizations. This method appeared in France by the end of the eighties and nowadays it is widely used in France, Spain, and other countries of Europe, as well as in Canada, and some Latinamerican Institutions. In France, ETED is promoted by CEREQ (Céntre d´Etudes et de Recherches sur les Qualifications. Center of Studies and Research about Qualifications) (Irigoin and Vargas, 2002); to an international scale there is a net that tackles it, which is named ETED Net, research and Development.

In this method a great importance is given as to Individual part in the tackling of an employment as to sociological and structural analysis of employment; by considering its relationship with other employments and with the more globalized economic processes through collaborative work among unions, social organizations and state entities. Moreover, it has into account the way how employments evolve, with the objective of making competencies and the management of human talent more suitable (Mandon and Liaroutzoz, 1999; Mandon and Sulzer 2002).

The ETED Method makes part of the constructivist approach of competencies, due to the fact that:

a. It conceives competencies in an integral form as movements or performances to reach certain purposes, in the area of relationship with the context (Irigoin and vargas, 2002). It does not center in a detailed analysis of attributes of competencies as it happens in the behavioral or functionalist approach (DACUM-AMOD-SCI and functional analysis).

b. It states the importance of analysis and reflection about work
c. It takes into account the relationship among occupations and professions as well as their purposes. It considers the social context too.
d. In the tackling of competencies, importance of variability is given, and there is consideration to the changes in the occupations.
e. It tackles dysfunctions in the context which must be solved with the competencies of people.

Some essential steps in ETED are (Irigoin and Vargas, 2002):

1. Identification of actors.
2. Interviews
3. Structuration of information
4. Writing and validity to the technical groups formed by the title holders of the employments

Among some critics, that this method has, are:

a. Although it sets a series of new aspects towards the former methods, such as the integration of dynamics of work and its variability and evolution, it tends to emphasize In the occupations (laboral), by considering a little to social, political, environmental and scientific aspects, or to personal development in the identification and formulation of the competencies.
b. It centers in the cards to describe the competencies, but it is not a systematized method to design programs of formation as such, which is an urgent need nowadays.
c. There is not a validated methodology of a shop, therefore, it is not systematizing to constructive work as it happens in DACUM method and in Focus group.

Focus Group Shop

This method, emerged in the qualitative research, and in the social and entrepreneurial organizations. It consists on making a participative shop in which participants are interviewed about the needs of the context and the competencies that the graduated must possess in a determined academic program. In this sense, there is work with a clear and defined focus to administer the time well and to impact in what is searched. In the ambiance of the construction of educative projects, the focus group is usually used with one, two or three reunions with experts in the field, students, professors, etc. The most common steps in this shop are the following ones:

1. Election of the shop coordinator.
2. Identification of the program which is going to be built by competencies.
3. Determination of the possible participants
4. Call to the participants
5. Explanation of the shop´s focus and/of the methodology of work
6. Generation of a confident environment in the group
7. Setting of questions to the attendants about the needs and requirements of the context.
8. Determination of the competencies.
9. Revision and validation of the competencies.

Here are some of the weaknesses or problems of the focus group method:

a. It tends to tackle very little to the components of the competencies, such as elements, criteria, knowledge, etcetera.
b. It does not consider too much to the contributions of the attendants to construct the processes of formation and evaluation. In this aspect, the education must see more towards the society and not to stay with its own perspective, because this last contributes to keep the same educative, traditional practices.

Reflexive Constructive Shop (Taller Reflexivo Constructivo: TRC)

This shop is a methodology of construction of programs of formation in all the educative levels. It is based on the contributions of the traditional methods of curricular design by competencies, but also incorporates the complex thinking, reflection about practice, educative action research and the academic quality management.

The reflexive constructive shop is applied to build the curriculum by competencies in all its components until reaching the establishment of the formative projects, through a clear action plan, with criteria, activities and chronograms. It is done through the observation and self reflection of curricular practices and micro curricular individual practices, the socialization, group reflection and collaborative construction. For this, group and individual practices are done once a week, every fifteen days or every month. It is done through the assignation of coordination roles, quality management and management of information.

In the frame 2.4 are compared the main characteristics of the Reflective Constructive Shop with the traditional methods, such as DACUM, the Functional Analysis, ETED and Focus Group. It can be seen that the Reflexive Constructive Shop is focused, since the first moment in building the structure of the academic project of the program in its diverse components. Furthermore, in the Reflexive Constructive Shop some suggestions of the participants are gathered to establish how the work should be done with the modules, the credits, the propaedeutic cycles and the policies of accreditation of the competencies, among other aspects.

an

Frame 2.4 Central elements of the constructive reflexive shop (TRC) and its differences with other methods.

Aspect	DACUM method	Functional Analysis	ETED method	Focal Shop (Or focus group)	Constructive-reflexive shop (TRC)
Orientation	To determine functions and tasks.	To identify the key purpose and the functions in a more detailed form from the context.	To analyze the processes, occupations in their organizative, relational and evolutive dynamics.	To determine needs, problems, lacks and areas of development.	To build the project of the program with its graduation profile, entry profile and curricular map from the analysis of problems.
Curriculum	To determine abilities, knowledge, and attitudes.	To build or to help to complement the graduation profile through units of competence and elements of competence.	It helps to identify the competencies of the graduation profile.	It helps to identify the competencies of the graduation profile.	To determine the components of the competencies according to the socioformative approach. It uses the educative action research.
Method	Group shop with 8-12 experts in the area.	Meetings with experts (compound groups by 5-10 experts in average.	· shops or enquiries to experts. · observation of the surroundings. · Analysis of the occupations in their relationship with the organizational context.	Group shop with several sectors. Students, professors, graduated and experts.	Group shop highly participative, in person or virtual, with experts and representatives of several sectors. Participants: Students, professors, graduated, experts of the area, etc.

Length	4-8 hours	Session: 2-8 hours. Average time of design of a functional analysis for a carrier of pre-grade: 1 year.	2-8 hours. In average, it lasts from six months to two years.	A session of 8 hours in average (only to identify the command of competence)	Twelve sessions in average, from two to four hours each one, until having the study of the referents of the program ready, the graduation profile, the curricular map, and the essential structure of the modules or formative projects. This is shared along a semester or a year.
Mechanics of the work:	Experts are requested about which general functions they make; this is written on paper sheets, on a board, and after, for each function the necessary activities, the knowledge, the abilities and attitudes are described.	-A moderator orients all the process. -some experts who know very well the area for the program are selected- A training time about methodology of work is done.	-one or two coordinators. -the laboral and professional dynamics are studied in their evolution and systemic relationship with the organizative context. -the profiles of competence are determined.	The representatives of the diverse sectors related with the program are investigated by the context needs and for the required competences.	-establishment of the leader of the project. -identification and call of the participants. -constructions of an action plan. According to the norms of work. -assignment of roles to favour the group dynamic, as the roll of management quality, the roll of manager of knowledge, and the roll of systematization of the achievement of the shop.

Aspect	DACUM method	Functional Analysis	ETED method	Focal Shop (0r focus group)	Constructive-reflexive shop (TRC)
	Phases: 1. Planning of the DACUM shop. 2. Execution of the shop. 3. Description of the competences. 4. Validation of the competences. 5. Final revision 6. Publishing.	· Experts are asked about the key purpose and the units of competence. After the key purpose and the units of competence are agreed, the elements of competence with the contributions of experts are identified. · Competences are described through the map of functional analysis.			· In the sessions each one of the components of the educative project of the program is tackled, and the contributions of the attendants are organized on a board or screen so that all are visualized. · Brainstorming work is done, co evaluation and continuous revision of the products during al the process. What is done is constantly published and everybody reviews it. · Before each session, in an individual form or by sub groups, some basic activities are done, according to the action plan.

ESSENTIAL PROCESSES OF CURRICULAR MANAGEMENT BY Competencies FROM THE SOCIOFORMATIVE APPROACH

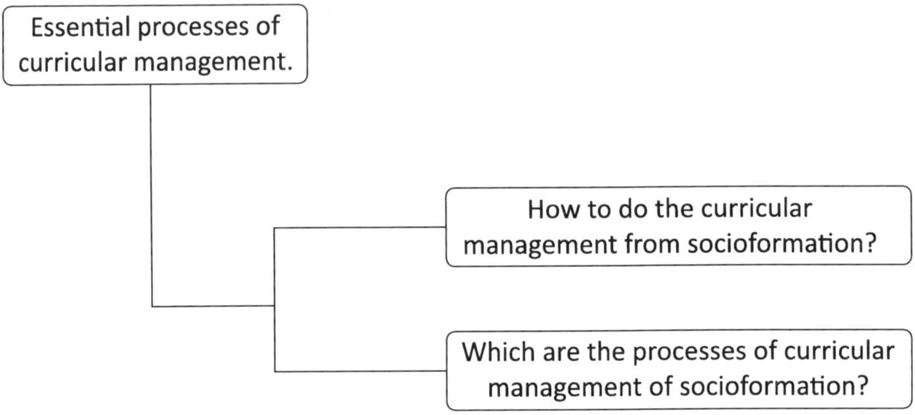

How to do the curricular management from socioformation?

Several guidelines and models to manage the curriculum by competencies have been proposed, as in basic and middle education as in superior education. (see, for example, Maldonado, 2001, 2006, Zabalza, 2003). From the socioformative approach it has been worked with the methodology described by Tobón, (2005) with nine axis which orient how to innovate the processes of formation with base in the employment of the strategies of complex thinking, the educative action research and the fifth discipline, of Peter Senge, (1995) (see frame 2.5).

This methodology of curricular management has been fully or partially implemented in several countries of Latin America and Europe (Spain and Portugal), and it has shown that it has impact to generate commitment by the teachers. It reduces the resistance towards change, it makes flexibility possible in the formation and evaluation of competencies and gives a higher commitment with innovation. However, the experiences of application of this methodology for almost 8 years in institutions of basic, middle and higher education have led to do an alternative systematization of the axis. It incorporates contributions of the models of management of excellence in the field of integral formation and of different tools to ensure the quality. This has favoured the configuration of a new model of curricular management which is called GesFOC (SysematicManagement of the formation of Competencies).

Most part of the contents of the contents of the nine axis of the socioformative curriculum (frame 2.5) have been integrated in this new proposal and have been complemented with new referents such as planning of leadership, the building and/or improvement of the educative model and the human talent management, the establishment of policies of formation and evaluation and the elaboration of the quality management of curriculum. In the figure 2.6 are exposed the essential processes of the methodology and its systemic relationship.

Then, it is proposed to consider the GesFOC model in all the curricular and micro curricular processes, as a key tool so that the actions of governments, states and society, on one side and the actions of the principals, administrative personnel, family and professors on the other side, are concatenated among them and they get to impact in the fundamental purpose: people to be self-fulfilled with an ethical commitment and they to have the necessary competencies to perform and occupation or profession, and to contribute to the social tissue, and to search the economic development, to generate science and technology, to give the socioeconomical development, and to participate creatively to culture, arts and recreation. Thus, the learning concept will be transcended and a new systemic work will be generated in which, the students will be principal actors of their life and of their own education.

The GesFOC model integrates several processes that are separately tackled in the educative institutions.

1. The design and curricular management of the educative programs with respect to the study of the context, the profile graduation and the entry profile and the curricular map.
2. The courses, subjects and/or modules, as well as the learning sessions.
3. The formation based in competencies.
4. The management and the insurance of quality of the educative processes through determined norms, as ISO norms.
5. The management of excellence in the organizational area.
6. The circle of quality in the academic management.
7. The management of human talent.
8. The evaluation of the educative impact in the formation and development of competencies.

The challenge is to tackle education from a perspective each time more systemic and complex, which allow us to understand the different dimensions of the human formation and all the factors of environment which influence in it, such as the educative institutions, the family, the society, the social organizations, the enterprises, the communicative means and the government. The GesFOC model allows this, because it considers the human development in its multi dimensions and it integrates the several factors related to education, in the frame of competencies.

It is necessary to state that with GesFOC model it is possible to manage the formation in several levels of analysis with the same factors, without the need of having other models. In the practice, this means that some international entities which work educative policies (such as UNESCO) can follow it, the educative ministries,

Frame 2.5 Axis of the socioformative approach

AXIS	ESSENTIAL ASPECTS
1. Educative action-research	· Observation. · Deconstruction. · Reconstruction. · Application-evaluation.
2. Self-reflection	· Self-evaluation. · Self-critics. · Metanoia.
3. Deconstruction	· Critical analysis of the institution´s curriculum. · Analysis of emptiness. · Determination of strength. · Analysis based on the seven knowledge of complex thinking. · Consciousness of mental negative models in curricular design.
4. research of the context	· Laboral-entrepreneurial requirements. · Social requirements. · Personal requirements.
5. Facing uncertainty	· Consciousness of uncertainty in the curriculum. · Setting of strategies to face the uncertainty in the competency-based formation.
6. identification of competences	· Identification of global competencies, of the units of competence and of the elements of competence. · Criteria of performance. · Essential knowledge. · Application rank. · Required evidences. · Problems. · Detection of uncertainty with the competency based performance.
7. Problematic nodes	· Global competence · General problems · Academic knowledge · Non-academic knowledge · Trans- discipline · Directed by teaching teams
8. Teaching team	· It directs a problematic, specific node · It has common objectives · Shared vision · Lines of research
9. Formative projects	· Formative route · Design of the project · Content of reference · Articulation with TIC

the educative institutions, and the teachers in the classroom, (presential or virtual) with its students. Obviously, it depends on whom it employs, the finalities and the emphasis are going to vary. Thus, for example for a minister the educative policies, the human talent, the resources, and the last evaluation will be more important whereas, the emphasis of a teacher will be set in the mediation process of the integral human formation.

Which are the processes of curricula management from Socioformation?

The processes of the GesFOC model are tackled as interrelated components which conform a system (with information of entry, processing and exit information) which self regulates according to the objective of people to have an integral formation. Complex thinking invites us to tackle curricular management as an organization in which flow multiple perspectives, visions, tendencies and interests. This is not a static process, but dynamic. On the contrary, it will be difficult to fully understand its sense. According to figure 2.5, each one of the twelve central academic processes are tackled through four interrelated actions which are an adaptation of Deming Model, to education (Tobón, 2013a): Directioning. (Establishment of goals), Planning (the necessary activities to achieve goals are determined), performance (execution of activities), and evaluation (assessment of the activities done and the achieved goals). In this form, the achievement of goals is ensured and the continuous improvement.

Traditionally, the curricular management has had activities of planning, execution and evaluation, but this is not enough. A clear directioning of the processes is required, in such a way that they possess as clear goals of what they intend to achieve as punctual criteria of evaluation which help to have the best possible quality. It is also necessary that the activities are done in a continuous manner and to avoid spaced periods, as it uses to happen with frequency.

For some people who are in front of curricular management, at the beginning, it can be difficult because of the tendency which has prevailed of tackling curricular management in a fragmented manner. However, when in curricular management, the different components are related it is more simple that traditional approach centers in each part separately, because the fact of relating the elements among them permits a better understanding and it gives as a result an easier performance.

What to do so that the proposed processes of curricular management have coherence among them and are directed towards the same route? With frequency, this is a problem in the projects of design, reform and curricular transformation. To tackle it, the concept of equifinality must be taken into account. This consists on searching that the axis get integrated among them through their contribution to a common goal which is the shared vision of the curricular management process.

On the frame 2.6. are described in a synthetic form, on one side, the different processes of curricular management by competencies from the GesFOC model,

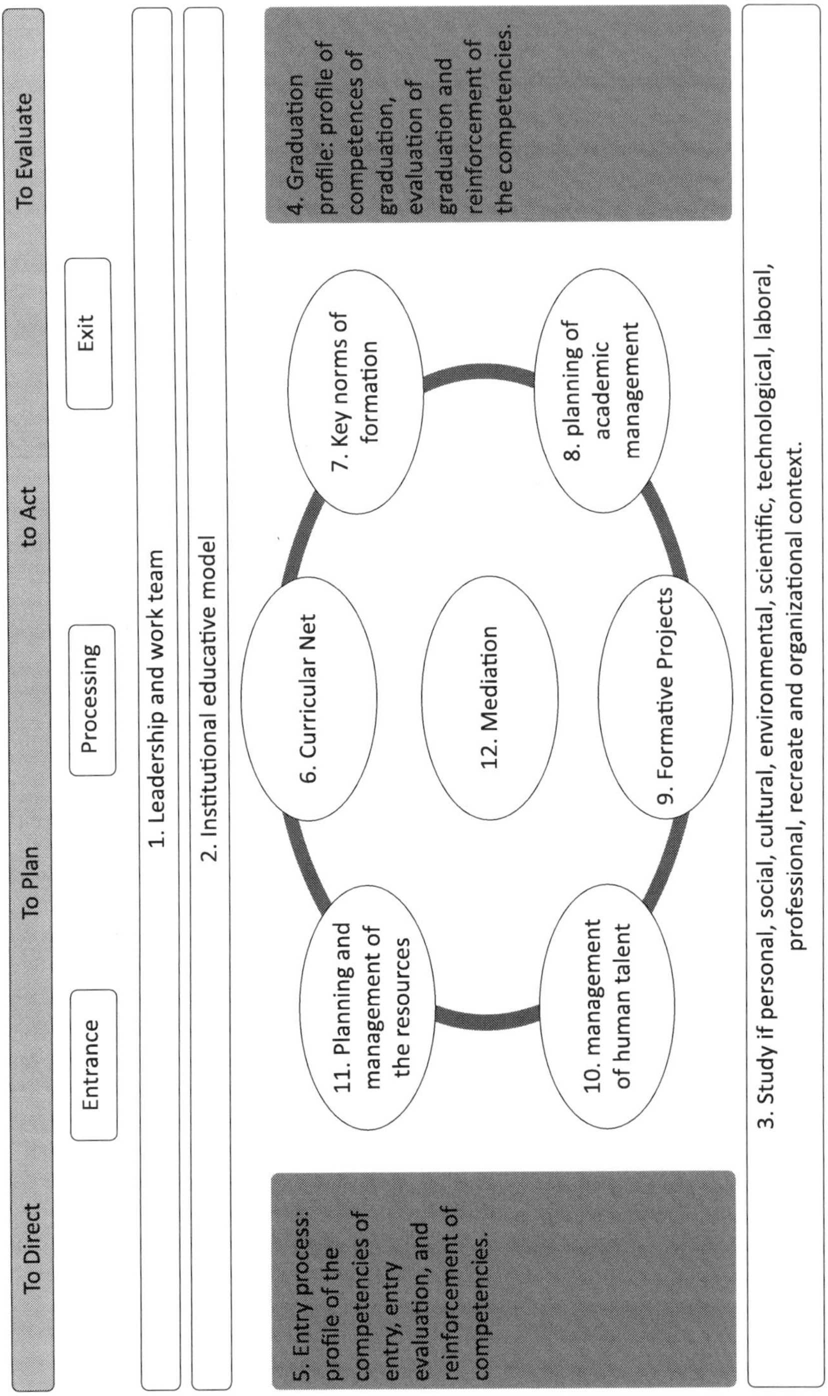

Figure. 2.6 structure of the GesFOC model to tackle curriculum.

Frame 2.6 Synthesis of the curricular management processes

1. Collaborative work and leadership	In every process of curricular management, a team which leads the processes and work in a collaborative manner is necessary, for searching actions to get to agreements about the fundamentals and to solve conflicts which may be present.
2. Educative model	In curricular management it is precise to construct and to improve an educative model, which orients the processes of curriculum, teaching, learning, anc evaluation from determined pedagogic referents.
3. Study of the internal and external context	The curricular management by competences must be supported on the continuous study of current and future challenges of internal context (institutional) and external (disciplinary, social, environmental, scientific-technical and laboral, professional).
4. Graduation profile	It is the set of competencies that the students must develop in the program, as the actions of reinforcement of such competencies and mechanism of help to graduated.
5. Entry process	It is the set of competences which the students must possess at the moment of entering a program of studies, and the actions of reinforcement which must be implemented in case of being necessary. This is a key to ensure the success in integral formation and the achievement of graduation profile.
6. Curricular net	This is the set of formative projects that the student must do along a program of studies. It has as a base the competencies of graduation and of entry, as well as the study of the context, the legal norms and the institutional policies.
7. Academic norms of formation and evaluation	They are the norms that directors, teachers, students must follow, concerning to management, admission, teaching, evaluation, scholarships and accreditation of studies.

8. Policies of Academic management	They are the agreed guidelines to carry out the processes of research , linking with the community and welfare of the students, teachers and principals. It considers the educative model, the graduation process and the curricular net.
9. Formative projects	They are the linked activities that students must do, mediated by the professors, with the objective of solving determined problems of the context, and thus, to achieve integral formation and the development of the expected competencies in the graduation profile.
10. Management of human talent	These are the necessary actions to search, to select, to form, to evaluate, to certify, to remunerate and to promote to principals, teachers, and administrative personnel with the objective of fulfilling with the educative model and the graduation profile.
11. Management of resources	They are the actions to identify and to adapt the required resources in the activities of formation and management, and to ensure to have them with the possible highest quality.
12. Mediation of integral formation and development of the expected competences.	They are the concrete actions that professors must do in the classroom with the students to ensure integral formation day after day and the development of expected competencies. This implies the students to get commited and to apply the learning strategies to ensure the achievement of goals.

The GesFOC model can be implemented in its totality by the same educative institution, or to look for having determined processes to be advised and helped by other institutions and organizations. For example, this could be helped by external experts to do the strategic planning, to design the curriculum and to plan the processes of formation. It could also count with independent organizations to carry out a certification on the ISO norms; and at last, it could be helped by organizations to manage its human talent and to evaluate to the students as it does Viveduc Center of Chile (Viveduc, 2009).

Bibliographic references

Deming, W.E. (1989). *Calidad, productividad y competitividad. La salida de la crisis [Quality, productivity and competitivity. The exit of the crisis]*. Madrid: Ediciones Díaz de Santos (edición original en inglés, 1982).

Domínguez, E. (2006). *Pensamiento complejo y educación. Una síntesis de las tesis de Morin [Complex thought and education. A synthesis of Morin's theses]*. En S. Tobón et al. (Eds.), *Diseño curricular por competencias [Curricular design by competencies]*. Medellín: Uniciencia.

Exiccom (2009). *Modelo de evaluación y certificación de competencias [Model for evaluation and certification of competencies]*. México: Exiccom (www.exiccom.org).

García Fraile, J.A., López Calva, J.M., López Rodríguez, N.M. y Aguilar Álvarez, A. (2012). *Gestión curricular por competencias en la educación media y superior. La perspectiva humanística del enfoque socioformativo [Curricular management by competences in secondary and higher education. The humanistic perspective of the socio-formative approach]*. México: GAFRA.

Irigoin, M.; y Vargas, F. (2002). *Competencia laboral: manual de conceptos, métodos y aplicaciones en el sector salud [Labor competence: manual of concepts, methods and applications in the health sector]*. Montevideo: Cinterfor-OPS.

Lev Vygotsky, L. S. (1985). *Pensamiento y Lenguaje [Thought and Language]*. Buenos Aires: Pléyade.

Maldonado, M.A. (2001). *Las competencias, una opción de vida [The competencies, a life option]*. Bogotá: ECOE.

Maldonado, M.A. (2006). *Genealogía de las competencias [Genealogy of competencies]*. Bogotá: ECOE.

Mandon, N., y Liaroutzos, O. (1999). *Análisis del empleo y las competencias: El método ETED [Analysis of employment and competencies: The DTS method]* . Buenos Aires: Libris S.R.L.

Mandon, N., y Sulzer, E. (2002). *Análisis del trabajo: La descripción de las competencias mediante un enfoque dinámico de los empleos (Versión electrónica) [Analysis of work: The description of skills through a dynamic approach to employment]*. Calificaciones & Empleo, Dimensiones francesas y europeas de la formación y el empleo - convenio Piette/Céreq, 33, 1-4. http://www.cereq.fr.

Martínez, S., Frade, L., Pimienta, J. y Tobón, S. (2011). *Evaluación de las competencias en pruebas masivas [Evaluation of competencies in massive tests]*. México: SEP y CREFAL. En prensa.

Mériot, S.A. (2005). *One or several models for competence descriptions: does it matter?*, Human Resource Development Quarterly, 16, 2, 285-292.

Mertens, L. (1997). *Competencia laboral: sistemas, surgimiento y modelos [Labor competency: systems, emergence and models]* . México: CONOCER-CINTRFOR/OIT.

Monereo, C. (2007). *Estrategias de enseñanza y aprendizaje [Teaching and learning strategies].* Barceona: Graó.

Morin, E. (1995). *Introducción al pensamiento complejo [Introduction to complex tought].* Barcelona: Gedisa.

Morin, E. (2002a). La mente bien ordenada *[The well-ordered mind].* Barcelona: Seix Barral.

Morin, E. (2002b). Los siete saberes necesarios para la educación del futuro *[The seven knowledges required for the future education].* Buenos Aires: Nueva Visión.

Norton, R. E. (1995). SCID: *A systematic competency-based curriculum development model.* Columbus: Ohio State University.

Norton, R. E. (1997). DACUM Handbook (2da. Ed.). *Columbus: Center on Education and Training for Employment,* The Ohio State University (revised 2008).

Pascal (1976). *Pensamientos. Texto producido por León Brunschwicg [Toughts. Text produced by León Brunschwicg].* Paris: Garnier-Flammarion.

Perrenoud, P. (2004a). *Diez nuevas competencias para enseñar [Ten new competencies for teaching].* Barcelona: Graó.

Perrenoud, P. (2004b). *Desarrollar la práctica reflexiva en el oficio de enseñar [Developing the reflective practice in the teaching office].* Barcelona: Graó.

Pimienta, J. (2012a). *Las competencias en la docencia universitaria. Preguntas frecuentes [Competencies in university teaching. Frequent questions].* Naucalpan, Estado de México: Pearson – Prentice Hall.

Pimienta, J. (2012b). *Estrategias de enseñanza-aprendizaje. Docencia universitaria basada en competencias [Strategies of teaching-learning. University teaching based on competencies].* México: Pearson.

Senge, P. (1995). *La Quinta Disciplina en la Práctica [The Fifth Discipline in Practice].* Barcelona: Granica.

Tobón, S. (2001). *Aprender a emprender, un enfoque curricular [Learning to undertake, a curricular approach].* Medellín: Funorie.

Tobón, S. (2005). *Formación basada en competencias. Pensamiento complejo, diseño curricular y didáctica [Competence-based formation. Complex thought, curricular design and didactic]* (2da. Ed.). Bogotá: ECOE.

Tobón, S. (2006). *Las competencias en la educación superior. Políticas hacia la calidad [Competencies in higher education. Policies towards quality].* Bogotá: Ecoe.

Tobón, S. (2009a). *La formación humana integral desde el proyecto ético de vida y el enfoque de las competencias [Integral human formation from the ethical project of life and the competencies approach].* En E. J. Cabrera (Ed.), *Las competencias en educación básica: un cambio hacia la reforma [The competencies in basic education: a change towards reform].* México: Secretaría de Educación Pública.

Tobón, S. (2009b). *Proyectos formativos: didáctica y evaluación de competencias [Formative projects: didactics and evaluation of competences].* En E. J. Cabrera (Ed.), *Las competencias en educación básica: un cambio hacia la reforma [The competencies in basic education: a change towards reform].* México: Secretaría de Educación Pública.

Tobón, S. (2010). Proyectos formativos: metodología para el desarrollo y evaluación de competencias *[Formative projects: metodology for the development and evaluation of competencies].* México: Book Mart.

Tobón, S. (2011a). *El currículo por competencias desde la socioformación. ¿Cómo podemos cambiar nuestras prácticas educativas para asegurar la formación integral? [The curriculum by competencies from the socio-formation. How can we change our educational practices to ensure integral formation?]* México: Limusa.

Tobón, S. (2011b). *La evaluación de las competencias por medio de mapas de aprendizaje: una propuesta frente a los métodos tradicionales de evaluación [The evaluation of competences through learning maps: a proposal against the traditional methods of evaluation]*. En Moya, J. y Luengo, F. (coords.), *Estrategias de cambio para mejorar el currículum escolar: Hacia una guía de desarrollo curricular de las competencias básicas [Strategies of change to improve the school curriculum: Towards a curriculum development guide of the basic competences]* (pp. 135-149). Madrid: Ministerio de Educación. Disponible en: http://www.cife.ws/comunidad

Tobón, S. (2011c). *Manual sintético de gestión del talento humano por competencias y la Quinta Disciplina [Synthetic manual of human talent management by competencies and the Fifth Discipline]*. México: CIFE.

Tobón, S. (2013a). *Formación integral y competencias. Pensamiento complejo, currículo, didáctica y evaluación [Integral formation and competencies. Complex thought, curriculum, didactics and evaluation]*. (4ta. Ed.). Bogotá: Ecoe.

obón, S. (2013b). *Evaluación de las competencias en la educación básica [Evaluation of competencies in basic education]*. (2 ed.). México: Santillana.

Tobón, S. (2013c). *Planes y programas de estudio por ciclos propedéuticos y competencias. Hacia un nuevo paradigma curricular desde la socioformación [Plans and study programs by propaedeutic cycles and competencies. Towards a new curricular paradigm from socio-formation]*. México: Trillas.

Tobón, S. y García Fraile, J.A. (2006). *La formación de las competencias en la educación superior [The formation of competencies in higher education]*. Madrid: Universidad Complutense de Madrid.

Tobón, S. y Mucharraz, G. (2010). *¿Cómo aplicar las competencias en la práctica docente? [How to apply the competencies in the teaching practice?]*. México: Kundrivi.

Tobón, S., García Fraile, J.A., Carretero, M. y Rial, A. (2006). *Competencias, calidad y educación superior [Competencies, quality and higher education]*. Bogotá: Magisterio.

Tobón, S., Pimienta, J., y García Fraile, J.A. (2010). *Secuencias didácticas: aprendizaje y evaluación de competencias [Didactical sequencies: learning and evaluation of competencies]*. México: Pearson.

Tobón, Sergio (2002). *Modelo pedagógico basado en competencias [Pedagogical model based on competencies]*. Medellín: Funorie.

Tobón, Sergio (2004). *Formación basada en competencias. Pensamiento complejo, diseño curricular y didáctica [Formation based on competencies. Complex thought, curricular design and didactics]. (1era. Ed.). Bogotá: ECOE.*

Viveduc (2009). *Plan de evaluación de las competencias en diferentes profesionales y niveles educativos [Evaluation plan of competencies in different professionals and educative levels]*. Chile: Viveduc (www.viveduc.cl).

Zabalza, L.A. (2003). *Competencias docentes del profesorado universitario. Calidad y desarrollo profesional [Teaching competencies of universitary teachers. Quality and professional development]*. Madrid: Narcea.

Zúñiga, L.E. (2003). *Metodología para la elaboración de normas de competencia laboral [Methodology for the elaboration of labor competency standards]*. Bogotá: SENA.

CHAPTER 3

Design of the curriculum from the Socioformation

How to do reality the curriculum by competencies and not to stay in a theoretical discourse?

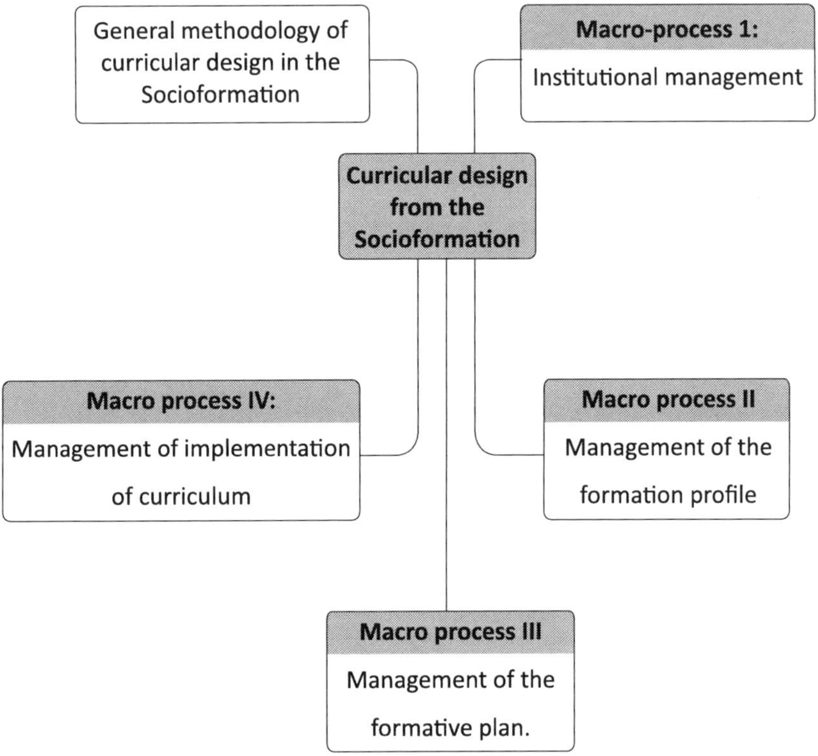

Figure 3.1 Essential axis in the curricular design from Socioformation.

GENERAL METHODOLOGY OF CURRICULAR DESIGN FROM THE SOCIOFORMATION

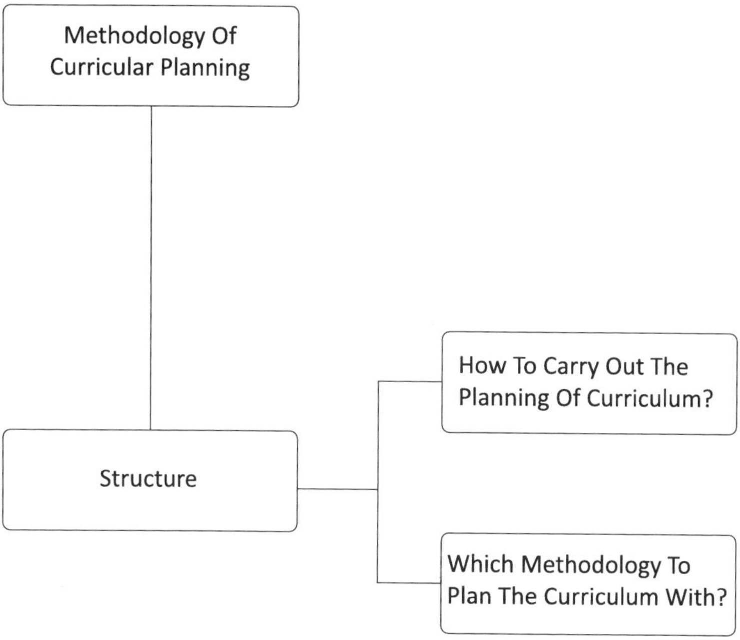

How to carry out the planning of curriculum?

The planning of the curriculum in the different educative levels is carried out by taking as reference the GesFOC model (figure 3.2), with emphasis in 12 processes which go from leadership until mediation. In this form, as the direction is considered as planning and its action with the students. In other words, The Macrocurriculum, the Meso Curriculum and the Micro Curriculum.

In difference to traditional methodologies of curricular planning–which remain in the construction of the studies plan-, the curriculum tackled form GesFOC model, assumes the human, integral formation, as a project in a continuous construction and improvement. Such a project is done in a participative manner In a determined educative institution, and it considers the essential processes to achieve the educative goals, as it is the case of team work, the educative model, the quality assurance, the management of talent and the evaluation of results.

Therefore, to plan the curriculum by competencies, it is not enough to do a study of the context, to have a graduation profile, to do a curricula map, or to plan the formative spaces. It is also necessary to do this in the frame of a more complete model of management and quality

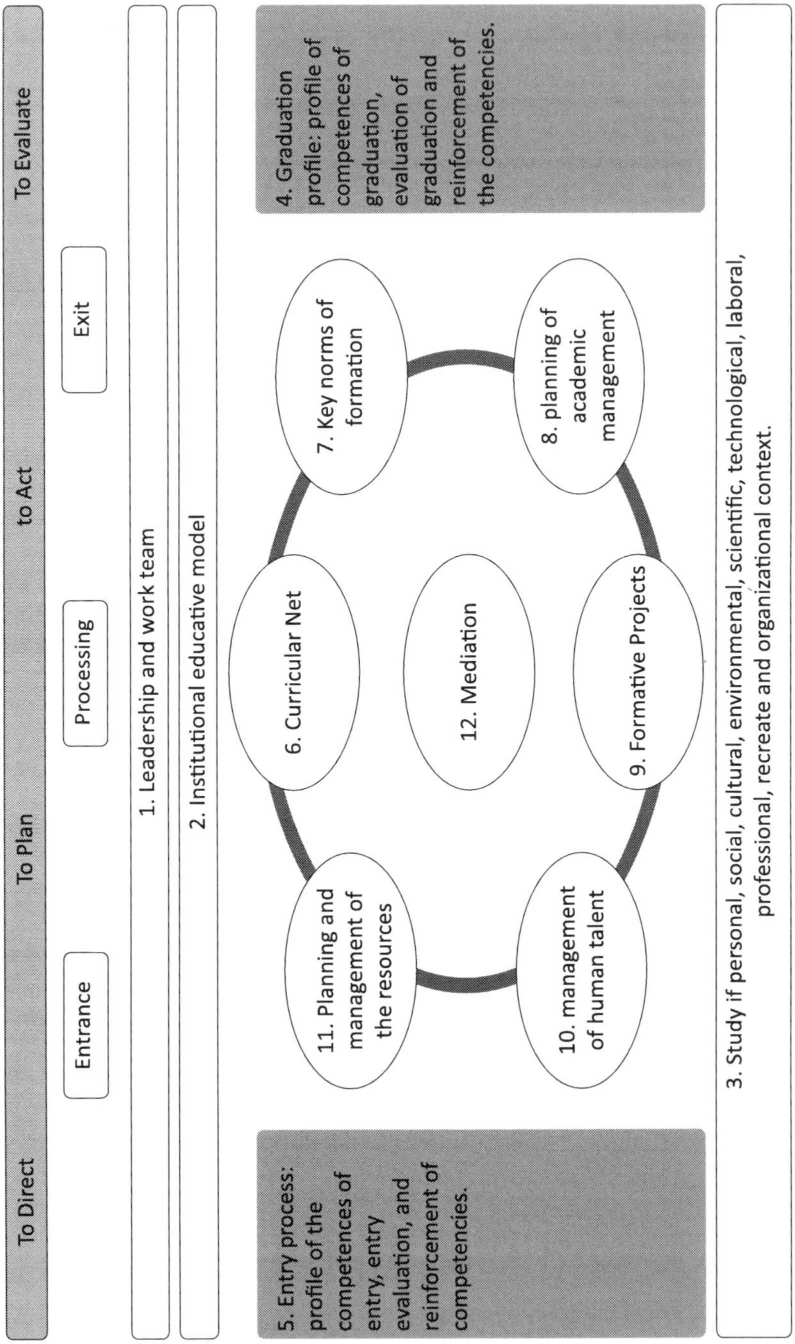

Figure 3.2 Components of the GesFOC model, to tackle curriculum.

assurance, which permits its continuous improvement after having set it in practice. This is one of the major learning that come after having several experiences about curricular design by competencies. It has been possible to corroborate that the lack of quality assurance diminishes in a notable form the impact of this approach.

Before beginning the description, it is important to set a warning: when you see the model, maybe it can be thought that there are many the processes to be tackled, however, in the nowadays bibliography the fact of planning a curriculum by competencies tends to be tackled, before all, by establishing a profile, designing a net and then, to plan the modules. Although in the GesFOC model, the processes are many more than the established ones in the specialized academic bibliography, the processes are simple and the focus in the essential. Besides, the application of this model has not set more time than the traditionally dedicated to the construction of a curriculum, which is from one to four years.

In the curricular planning, it is important to begin with a general evaluation of the estate in which the institution is or the program in each one of the 12 processes of the model. Then, it is important to plan the activities to be implemented in the required processes. At last, to do the planned activities , to verify the reached goals, to take actions of improvement, and to document –to socialize what has been elaborated.

To proceed in this form, the achieved goals are retaken and it does not begin from zero. This is a common practice in the nowadays curricular design because it tends to consider the past as inappropriate or non-pertinent without any serious analysis.

Besides, this generates an increase of time of curricular design as it is in the investment of human talent. The purpose is, therefore, to start from what the institution possess and from there, to propose activities which innovate the formation.

Therefore, it is not necessary to plan a curriculum with the 12 minimal processes of GesFOC model, the processes to be tackled depend f the starting evaluation. This is made, whether in the institution as such or in the program or in the course (subject, module or formative project): To achieve this, the guidelines established in the diagnostic must be considered and in the criteria which are in each one of the processes.

Quality assurance: this the process which helps to give warranties to the students, to the institutions, supervisors of the educative service and to the community, that the educative project is pertinent –according to the established educative policies- and it considers the real current and future social, laboral-professional, environmental, scientific, cultural challenges, etc. Even more, in this warranty, the institution establishes that it gives the formative mediation through suitable professors and principals, with learning strategies and of evaluation, coherent with the goals of formation and the challenges of the context, as well as the appropriate resources to such strategies and goals.

An institution may warranty the educative quality if first it manages the quality in all the processes (Tobón, et al, competencies, 2006), as the GesFOC model proposes, for example, and second, it must offer tangible and documental evidences that it possess the minimal, required quality so that the students reach the goals of formation. To do this last, it is necessary to have internal and external auditors which validate such quality and, if possible, to have such quality of academic processes to be certified (for example of ISO 9000 norms). It is also relevant that the human talent certifies its suitability with an independent institution, in such a form to have an evidence that the professors, principals and administrative personnel possess the required competencies to occupy their jobs.

Which methodology to plan the curriculum with?

The methodology to plan the curriculum by competencies consists on tackling each one of the 12 minimum processes of GesFOC model and to work the four phases of the figure 3.3 through continuous improvement. These four phases integrate the cycle of total quality of Deming (1989) (to plan, to do, to verify and to act), as well as two very important axis to assure the educative change: the directioning and the documentation-socialization of the actions (Tobón, 2010).

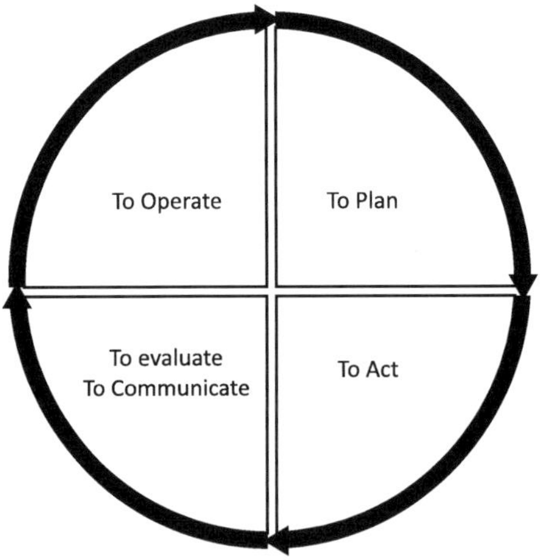

Figure: 3.3 Cycle of quality in the GesFOC model.

To Deming´s cycle of Quality has been added the operation at the beginning because it has been seen that in education it is necessary to emphasize that all planning must be done from this vision, some appropriate criteria and a diagnostic. Although we consider that Deming´s planning phase sets out the diagnostic, in education, this can not remain only in the external form, but it must take to every member of the educative institution to watch and to reflects upon his practices to make consciousness of the need of improvement This connects with the thoughtful approach of Schon (1992, 1998), and the practice of educative action-research (Elliot, 1994). Besides, in the management of educative quality it is insisted a lot to begin the processes of management for self-evaluation, which is considered as fundamentals in the processes of accreditation of the universities.

In the addressing are identified and/or built the criteria and of each process of GesFOC model to orient the starting. It is intended that people do a self reflection about how they are orienting the management of formation according to the considered criteria, therefore, the field diary , tests are appropriate. This allows to notice the achievement, emptiness, insufficiencies, negative mental models, and processes of simple thought (rigidity, resistance to change and fragmentation of teaching) (Tobón, 2009 a, 2010 a, 2001 a, d).

With respect to the documentation-socialization, CIFE´s experiences show that many times the educative institutions have an appropriate and solid academic, quality management but they do not have any documentation of what they do and with this, it is very difficult for them to show the impact they achieve. Likewise, it is very common to observe that the institutions only document what they do when they are going to receive a visit of an external auditpry or of pairs, because they tend to ignore or to forget it. Therefore, in the evaluation, the documentation is important .

1. To address: it consists on determining the products and criteria which will be taken into account in each one of the processes of curricular management. These are determined by considering the educative policies established in the international national and institutional levels. Based on these criteria, a diagnostic about achieved advances, in the corresponding process, can be made and to take consciousness of the curricular practices to generate actions of change.

2. To plan: it consists on determining the activities which will be done in each one of the processes of curricular management, by following determined patterns of the socioformative approach, and by considering the products and criteria of addressing.

3. To act: once planned each curricular process, it is executed by following determined guidelines, whether by the team of curricular management or by each professor. In the execution, it is precise o tackle unexpected problems.

4. To evaluate: it is essential to to a continuous evaluation of achievement of the criteria of each process. It is also recommended to document the done activities and the reached results, as well as the difficulties and the solutions for such difficulties. Likewise, it is precise to socialize the advance and the results of the different processes of curricular construction in the educative institution, as the changes in the teaching practices (frame, 3.1).

Frame 3.1. General actions for curricular management.

Actions	Suggestions
To address	· To determine the product or products to be achieved with their criteria. · To look for the criteria to be appropriate and to respond to the educative policies and to the educative model of the institution. · The criteria must be socialized, validated and/or appropriate for the team. · To promote the de-construction of the non appropriate teaching practices. · To do reflexive shops. · To do seminars and congresses. · To invite professors, principals and students to self-reflection. · To do a diagnostic about the advance in the respective process in the institution.
To plan	· To plan the activities to implement the process and to achieve the products and criteria for addressing. To take into account the diagnostic which has been made. · To look for the methodology to plan the process to do the most simple as possible, but without losing strictness. · To review the panning and to improve it, to be it appropriate to the institutional and /of the context challenges.
To act	· To execute the planned activities · Each certain time, to reflect upon how planning is being executed, to start improvement and changes in it. · To consider the possibility of introducing changes, although they do not agree with what was planned, if that contributes to achieve the established goals improved. · To train to people to execute the process as it is. · To generate key agreements to do reality the new curriculum and not to leave it only in the paper.
To evaluate	· To compare the activities which have been done, with what was planned and the criteria of reference, and to establish the grade of fulfillment. · To value the use of resources, of human talent, and of time, according with what was planned. · To identify difficulties and the manner to tackle them during the execution of the process. · To introduce changes in the methodology to improve the process if necessary. · To improve the management of resources, of human talent and of use of time. · To document the activities which have been done, and the fulfillment of the base criteria. · To socialize the activities with the team and with the educative institution.

MACRO PROCESS I: INSTITUTIONAL MANAGEMENT

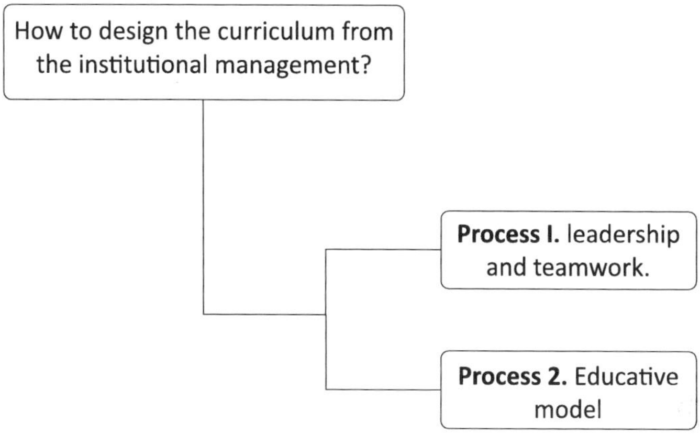

Process 1. Leadership and teamwork

Concept. the starting step to the management of the curriculum by competencies is to form a solid team which, on one side, leads the elaboration and /or revision of the educative model, and on other side, which orients the curricular management in the diverse processes described in the current chapter. Then, it is necessary the team to be legalized in the institution, its members to be have an adequate assignation of time , the team to be recognized by the academic institution and that in the team exist distribution of roles among the members.

The team or the teams that participate in the management of curriculum (and/or micro curriculum) must have a clear leadership to assure the work with an agreed action plan. The work team and the leadership must be evaluated regularly, to establish improvements. It is also necessary to generate a confident environment in the team to assure the success of the activities with different points of view and feedback sources. To achieve this, it is necessary each member to assume a committed attitude with the team, and to participate in the assigned activities by doing theoretical and practical contributions. This requires of flexible, open teams which give confidence to the members.

The team must establish an action plan with base on the diagnostic of the state that the institution has, the program or the micro-curriculum with respect to each one of the 12 minimum processes considered in the GesFOC model. It must take into account the criteria of each process, with the objective of establishing obtained achievements and the aspects which must be improved, changed, or created. In the action plan are prioritized determined processes and activities, and there are some periods established to execute the established tasks, as well as the required human talent, the form to have teamwork and the necessary resources.

To direct: Product: document with the action plan in turn to the design and/ or improvement of curriculum.

Criteria:

1. Leadership is formalized in the educative institution, according to the requirements.
2. There is teamwork in the academic and curricular management in the educative institution according to the institutional policies.

To Plan:

1. To agree how the team work is going to be formed in the institution so that it becomes representative.
2. To determine the activities of the team in the academic and curricular management.
3. To establish how the team´s coordination will be and to determine the rolls considered to be necessary.
4. To define an action plan for the team and to consider all the processes of curricular management.
5. To establish a schedule of meetings and the periods to be carried out.
6. To look together with the authorities the members of the team to have an assigned number of hours per week.
7. To design a formation plan in curriculum, competencies, didactics and evaluation.
8. To establish mechanisms to value the work in the curricular management as an academic product. To have into account in the process of pay back/and or in the promotion in the teaching principal category.

To Act:

1. To analyze the previous team work experiences and based on these, to contribute with the team to work out and to have continuity.
2. To do a group reflection of how to have into account the learning of previous team work in the educative institution, to prevent difficulties or problems and to assure the success.
3. To self reflect about possible, negative perceptions around the team work in the educative institution because of possible difficulties. And to assume the new team as an opportunity of having a positive experience that lead to concrete products of a high academic value.
4. To work in team on the frame of the established goals and to

accept the differences, by completing with the competencies of the members.

5. To execute the action plan with the distribution of responsibilities among the members of a team of curricular management, by following up a work timetable (see frame 3.2).
6. To carry out a training process of the members through phases, along with the application of what they learn.
7. To tackle uncertainty situations with a proactive attitude and a positive thought.
8. To determine actions, if necessary, so that the activities of the members of a team value them as relevant, academic products.

In the frame 3.3 (p. 156), there is a form which can help in the establishment of the action plan and the consolidation of leadership.

To evaluate-to-communicate

1. To evaluate how leadership and team work are.
2. To determine if decisions are taken in a participative form and according to the traced goals.
3. To establish if teamwork corresponds to the approved action plan and with the challenges of curricular management in the institution.
4. To verify if the team is sufficiently recognized in the institution and if the dedicated time of its members is in their monthly course load.
5. To do a regular register of the work process in team, because this labor is very important for the quality management.
6. To socialize the experience of the team work in the institution.

Roll	*Definition*	*Key Activities Of The Person*
Coordinator	The coordinator is who leads the teamwork	· To form the team, to look for experts and to lead the establishment of the action plan. · To explain the methodology in a clear and simple form. · To motivate to the participants to express their ideas without fear. · To create a confident environment where other´s ideas are respected. · To generate brainstorming of problems, competencies and guidelines of formation. · To look for agreements about the products to be achieved and about the action plan in the curricular management. · To orient the work of curricular management. · To give the word to the members of a team · To manage the organization of the information on a visible, graphic map for everybody, by using a computer and a projector. · To look for contributions of people to be complemented among them to generate group products. · Before contrary positions of the attendants, it focuses on common points to generate agreements and to advance.
Organizer	The organizer is in charged of documenting, keeping and getting back information about the activities and products of curricular design of the program.	· To write down the details of the contributions of all the experts and members, as well as of all the people who have some roll in the process. · To do graphic maps with all the generated agreements and to set out the points where there is disagreement. · To systematize the products with the mediation of the coordinator. · To register, in an official form, the products of the curricular management. · To present reports of the agreements. · At the end, to present the report of the agreements and of the points under discussion To ensure the protection of the products, faced to loses, damages and unauthorized uses.

Roll	Definition	Key Activities Of The Person
Manager of quality	Definition The manager of quality searches the product to have the expected characteristics, according to the educative model of the institution and the criteria of national and international accreditation.	· To deal with the four phases established in the GesFOC model: addressing, planning, performance and evaluation. · To evaluate the dynamics of the work and to do suggestions for its improvement. · To verify the quality of the products. · To afford suggestions to improve the activities and products of curricular management. · To present regular reports of the fulfilment of the action plans and the grade of achievement of the proposed goals in the criteria.
Coach	The coach is in charged of generating a good working environment and of implementing actions to overcome the possible resistances that might emerge.	· To implement actions to achieve the team work. · To afford suggestions to have a favorable environment in the teamwork. · To implement actions to overcome the resistances in the curricular management.
Participants in general	The contribute to the achievement of the products and criteria in each process.	· To do the agreed and assigned tasks. · To establish reflections around the competencies to for in the students and their approach. · To review the advances in the curricular construction and to afford suggestions for the improvement of the academic program.

Members of the team	*Key roles in the work team*

Products To Give In The Curricular Management	**Date Of Delivery Of The Products.**

Activities	Date to do the activities	Necessary Resources			
		Description Of The Resource	*There is*	*It is necessary to look for it*	*It is necessary to adapt it*

Process 2. Educative model

Concept. The team of curricular management is the one in charged of adapting or complementing the institutional educative model to the program, with its particularities. If there is not an institutional educative model, or this is does not match with the competency approach, it is necessary to promote its improvement or to work with the most general elements of the institutional philosophy.

An educative model is defined as a set of educative general guidelines which orient to the institutional community around what type of person to form, what society for, in which educative spaces, the didactic and evaluative guidelines to work with, what general philosophy, and what quality management model to work with (Tobón, 2006, 2010 a, 2011 a). From there, the design of an academic program requires of an educative model to have a solid fundamentals.

To address. Products: document with the educative mode adapted or contextualized to the program of studies.

Criteria:

1. Construction, revision and/or improvement of the educative model of the institution. To consider the educative challenges and the institutional philosophy.
2. Strategic planning, each certain period, with participation of the diverse academic areas.
3. Implementation of basic guidelines of formation based on competencies, in the educative model, in articulation with the challenges of the context and the current educative policies.
4. Socialization of the educative model, and to look for the appropriation in the academic community.

To plan:

1. To plan the basic aspects that this educative model is going to have, by considering the pointed referents in the frame 3.4 such as: type of person to form, teaching guidelines, learning and evaluation, curricular structure of the programs, methodology of the competencies, etc.
2. To determine, from the seven knowledge of complex thinking, which referents the educative model is going to have and how they will be tackled (Tobón, 2004a, 2009b, 2010 a, b).
3. To define the roles in the team work to construct and/or to review each one of the aspects that are expected in the educative model.
4. To identify if the team requires complementary formation to do the activities of the educative model and to manage such information.

To act:

1. To identify the implicit or explicit referents that the institutional educative model has had concerning to the institutional philosophy, teaching, research, extension, etc. and to know how it has been done in the practice. To detect emptiness and inconsistencies. From this, to formulate mechanisms to avoid to have similar situations in the future.
2. To determine if the current values, which orient the educative model of the institution agree with the expected seen from an institutional and society point of view. In case they do not, to determine the reason for this and to establish standards to eliminate disagreements.
3. To reflect upon the establishment of the seven knowledge of complex thinking in the implicit or explicit educative model of the institution (Morin, 2002, Tobón, 2010a)
4. To construct and or to review each one of the agreed aspects of the educative model with base on the assigned roles.
5. To address the problems which appear during the pro-active process and the positive thinking.
6. To look for the educative model to be easy and it to give concrete guidelines with respect to the educative organization, formation and value of learning.

Frame 3.4 Key components of the educative model.

Key aspects of the institutional educative model	*Description*
Which type of person is intended to be formed?	· To describe which type of person is intended to be formed in the institution and what his participation will be in the society, in the economy and in the environment.
What type of society is the person intended to be formed for?	· To describe the characteristics of the society the teachers, principals and students are going to be formed for. (the nowadays society and the future society might be described). · To indicate in general terms what type of connection with the society is expected in the educative institution and some possible strategies to achieve it.
Which educative approach or approaches is it intended to form?	· To describe in a synthetic form which is the educative approach or approaches which are going to orient in a general manner the formation in the institution. · To describe the concrete guidelines to consider them from an educative approach of reference.
Which educative principles is it intended to form with?	· To describe some basic guidelines to consider them in the learning, teaching and evaluation process, which orient to professors, principals and students. · To indicate some guidelines around some complementary courses of formation, as well as the paper of the extracurricular activities of formation.
Which is the roll of the teachers?	· To describe the general guidelines around the teacher roll in the formative process, based in the educative approach or approaches to be followed.
Which is the paper of the students in the formative process?	· To describe the general guidelines about the roll of the students in the formative process, by having as a base the educative approach or approaches to be followed. To consider the articulation of the activities that the students do with the teachers, with the independent and autonomous learning activities.

Key aspects of the institutional educative model	*Description*
Which sceneries is it intended to form in?	· To describe the educative and the guidelines modalities for each one of them: education, in person, virtual, distance, etc. To indicate if a program can have several kinds or it can not, and the path from one modality to another.
How must the basic structure be of the formation programs?	· To describe which basic curricular structure must the formation programs have to guarantee the meaningful learning of the competences. For example, if it is by subjects, modules, projects, propaedeutic cycles, etc. · To indicate the minimum aspects which have to be considered in the design of a formation program. · To describe the general guidelines which will be considered in the micro curriculum, with the objective of ensuring the competences of the students to form. · To describe the general model of quality management of curriculum and of the teaching process, by establishing, as minimum, periods of revision for the educative programs, how the evaluation of teaching quality will be, which referents to evaluate the quality with and who will evaluate it. · To determine in a general form the policies of promotion of the teaching, principal or administrative human talent, to consolidate an academic area. Here the concrete processes of management of human talent are not described in detail. This is left for the process 10, of curricular management.
How will the linking with the society be carried out?	· To describe in a general form the actions to carry out with the social, environmental, technological and entrepreneurial organizations. · To indicate how the actions of service to society are articulated, with the processes of formation of competencies.
Which will the basic premises be in the research?	· To present the key guidelines around the making of research in the program by the students, and teachers, looking for the articulation with the process of formation.

To evaluate-to communicate

1. To evaluate if the educative model is constructed based on the contributions of the educative community.
2. To determine if the institutional educative model corresponds to the institutional philosophy and with the norms and current educative policies in the country and in the region.
3. To determine if the institutional educative model has been evaluated by external pairs based on the institutional referents.
4. To verify if the educative model describes clearly the methodology by competencies which is intended to be followed in the educative institution, according to determined educative referents.
5. To establish if the educative model describes accurately the type of person to form, the general learning, teaching and evaluative policies and the general structure of curriculum of the academic programs.
6. To document the construction and improvement of the educative model.
7. To socialize the final document of the educative model with the educative community through several means, and to look for the community to give reflections and suggestions to improve and/or to apply such an educative model.

To construct an educative model it is necessary to begin by an institutional diagnostic and of the social, researching, and laboral-professional environment, to determine the educative challenges and to develop basic actions to face them with pertinence and creativity. Here is where the competency based approach offers a series of methodological orientations, to construct the educative model in such a way that it considers the challenges of the context, and in this form to make possible the development of educative, formative and appropriate programs. For this, it is necessary to understand that the educative model is the one which is effectively set on practice and it is not only set in the paper.

From the socioformative approach, it is recommended an educative model to have the suggested, components in the frame 3.4. As minimum, it is necessary to consider according to our criterion: the description of the type of person to be formed, the approach by competencies to be developed, the society for which the person is going to be formed, the philosophic principles to consider and the model of quality management which is going to be referent, in the self- evaluation.

In the frame 3.5 a synthesis of an example of an educative model is presented, from the components of the frame 3.4

Key aspects of the institutional educative model	*Description*
Which type of person is intended to be formed?	· To form professionals who help to generate a culture of health promotion in the community in order, to strengthen the quality of life of people. · To promote the full realization in people from their self- knowledge, based on the antropoetic commitment. · To form honest, integral professionals with an antropoetic commitment.
What type of society is the person intended to be formed for?	· It is expected to contribute to construct and to strengthen a self management and entrepreneurial society, with a high grade of commitment to have responsible citizens with quality of life, who are in conditions of promoting their mental health. This requires: -To promote the development and the strengthening of the social tissue, to generate a culture of peace and of coexistence, based on solidarity, teamwork, and cooperation. -To increase the scientific knowledge about the problems of mental health and its resolution to have a major impact in the processes of intervention.
Which educative approach or approaches is it intended to form?	· Complex thinking will be the base, and the Socioformation will be applied as In the studies plan as in the learning and evaluation practices, by considering the contributions of other pedagogical approaches. · Key guidelines: -To center in the ethical project of life of the students in each module or formative project. -to work with formative projects -to mediate so that the students develop competences which allow them to face the personal, social and professional challenges .

Which educative principles is it intended to form with?	· The students are searched to develop competencies through formative projects, by considering the 10 essential actions of the Socioformation.
Which is the roll of the teachers?	· In the program, the teachers are mediators to ensure the development of the competencies in the students.
Which is the paper of the students in the formative process?	· The students assume themselves as active and participant people, with an important challenge of constructing their ethical project of life. They develop the necessary competencies to face the current and future challenges of the context. They contribute to improve the actions of formation and evaluation and at the same time they get formed, they must give a service to others and to themselves, and or to the environment, to solve determined problems they meet.
Which sceneries is it intended to form in?	· The educative modality is mixed: - There are learning activities with direct help of the teachers and independent activities of the students. - The activities of independent work must be registered and be evident in products which are taken into account in the process of evaluation. - The learning ambiance must strengthen the participation and collaboration of the students.
How must the basic structure be of the formation programs?	· In the program, the curricular structure is going to be carried out through formative projects , in which the students must solve the problems of the context, with the application articulated of the know how to be, how to do, how to know and how to coexist. · The micro curriculum must have as minimum: - One or more competences to form. -Problem of the context -Criteria -Evidences -Essential activities -Key resources

Key aspects of the institutional educative model	*Description*
	· The model of quality management of the curriculum will be given by the implementation of four phases of GesFOC model: addressing, planning, acting, and evaluation-communication. · The management of human talent will be by competences, by following the same proposed referents with the students. This implies to select to principals, teachers, and administrative personnel, under this approach, to evaluate their performance through evidences and to promote their formation through projects oriented to the needs of development in each position. The 20% of compensation will have as a base the biannual performance from evaluation of the evidences.
How will the linking be carried out with the society ?	· It will be given through projects in which the students will do actions of promotion, prevention and intervention in mental health and management of human talent.
Which will the basic premises be in the research?	· In pre grade, the research will be of formative nature, although the students who have this interest will be able to do scientific research as such, going from projects with the teachers and researchers of the program. · Formative research could be done in the same projects of the studies plan or outside these.

MACROPROCESS II: MANAGEMENT OF THE PROFILE OF FORMATION

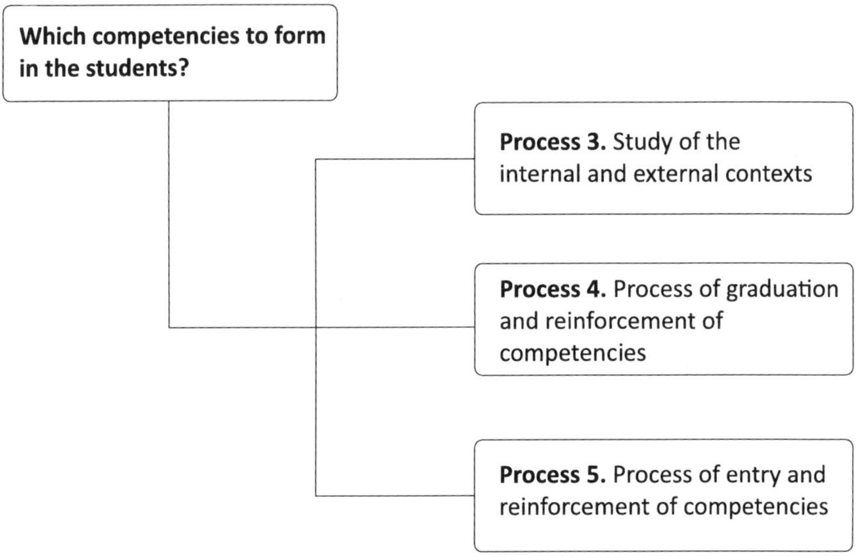

Process 3. Study of the internal and external contexts

Concept. The study of the internal and external contexts determines through different techniques the concrete challenges that the program will help to face in the students. This will be as in the present as in the future (form 5 to 10 years, approximately), having in mind the local, national and international field. Likewise, it describes the existence of real opportunities of employment or of entrepreneurship by considering the occupational tendencies. In this form, there is a report of the most important elements which support the importance and suitability of the program.

To form with base on competencies requires to have a high grade of knowledge and understanding of the social, laboral, political, economic, professional, entrepreneurial, and disciplinary challenges and requirements, together with the characteristics that human talent must have which is proposed to form by the educative institution. Such requirements must be established through systematic studies which provide exact information to be integrated to curricular design and to identification of competencies to form in the students, teachers, principals, and administrative personnel. The traditional curriculum has been done mainly from the academic to the disciplinary, but lacking the socioeconomical context.

The study of the context must be done as any diagnostic is done swiftly with agility, practicality and in the less possible time. When an institution takes long in this study (more than six months) it is possible that when they finish it, the context has changed too and this would mean to update such study.

The traditional approaches of competencies set out specialized methodologies to do the study of the context, as it is the case of the functionalist approach. In the Socioformation there is not a proposal of specialized methodologies, but to apply the common tools in the work with projects to do this type of studies, trying them to be suitable. This is what is proposed in the reflexive constructive shop (TRC), which speeds up the process and allows more participation of principals, teachers and students.

In the TRC it is proposed to consider the diagram of development which is exposed in the figure 3.4. Essentially, it tries to identify the challenges which must be under conditions of facing the students, principals and teachers of the program, as in an institutional level as in the social context. It starts with a revision of all the possible documentation about it (analysis of secondary sources). This is complemented with interviews to experts (from three to six).

If the revision of the secondary sources and the interview to experts is not enough to achieve a relevant study of the internal and external contexts, then it can be possible to go to interviews to a major number of people, by including, over all to employers, researchers, entrepreneurs, visionaries, students, graduated and teachers. If this continuous to be insufficient, then tests or questionnaires can be applied to people´samples.

The study of the internal and external contexts are led by the team of curricular management of the program, through practical shops, the collaborative work and the participation of different educative actors and the socioeconomical context.

Study of the internal context

It consists on identifying the problems which are précised to be solved in the program of current formation.

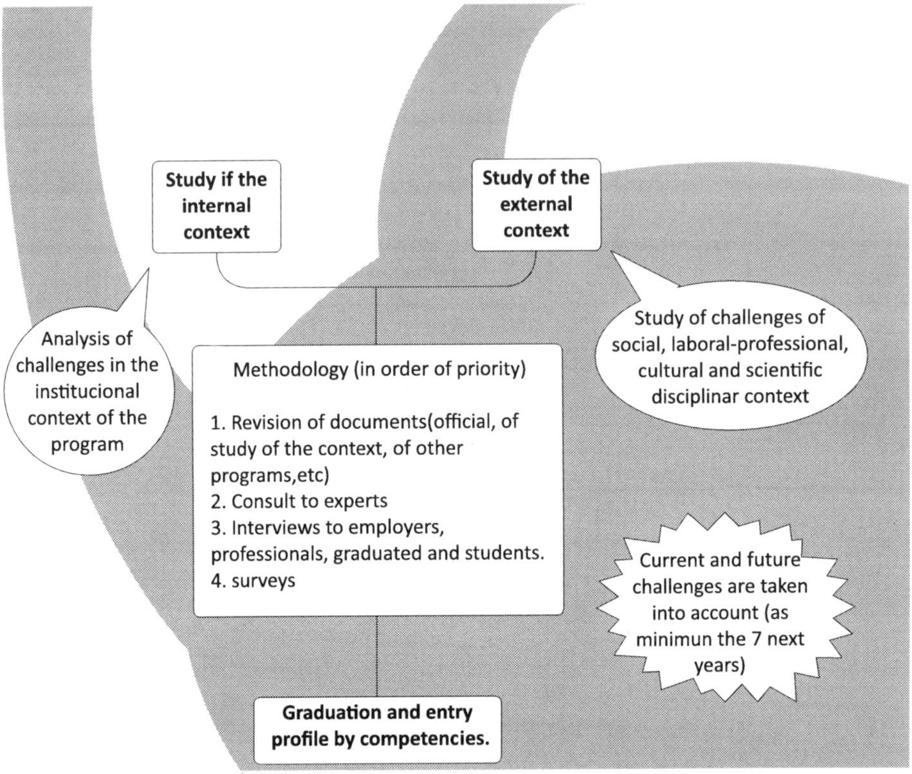

If the program is not yet created, the study of the internal context is not obligatory. However, with the objective of having elements to construct the curricular net, some general factors in other programs that the institution may have can be studied, if there are programs.

To address. Product: reports of studies of the internal contexts.

Criteria:

1. The challenges of the internal context are identified according to the characteristics of the institution.
2. The challenges of the internal context are determined, based on backed up evidences.
3. The challenges of the internal context identified can contribute to increase the quality of the programs of studies.

To plan:

1. Level of dropping out of school of students and teachers.
2. Level of graduation with qualification.
3. Level of graduation without qualification.

4. Average of academic efficiency
5. Subjects or modules with higher rates of disapproval index
6. Rate of employment in graduated
7. Average time that the students last to get their first employment.
8. Areas, subjects, or elements of the current program which increase employability
9. Areas, subjects or elements which are perceived as of little value for the formation.
10. Satisfaction of the students with the current program.
11. Satisfaction of the graduated ones with the current program.
12. Satisfaction of the teachers with the current program.
13. Satisfaction of the principals with the current program
14. Recommendations of the students, teachers, principals and graduated ones to improve the program.

To act:

1. To tackle the problems which get present during the process of pro activity and positive thought.
2. To look for the necessary documents to do a study of the internal context.
3. If the analysis of documents is not enough, to interview to people who have a deep knowledge of the institution and the program
4. To determine the challenges which must be faced in the program to achieve the maximum educative quality.
5. To establish, in a general form, the actions to tackle the challenges which the study of the internal contexts sets out.

To evaluate-to communicate:

1. To verify that there were the necessary sources for the study of the context, according to the criteria.
2. To verify that the found and systematized information tackle the established criteria.
3. To value if the challenges which were determined correspond to the information obtained from the consulted sources.
4. To value if the proposed actions to tackle the challenges in the program are feasible to implement.
5. To document the process of study of the internal context, as in the level of the followed process as in the achieved results.
6. To socialize the final document of the study of the internal context with the educative community through different sources and to look for the community to give reflections and suggestions to improve these aspects.

Study of the external context

Concept. The study of the external context consists on determining the problems (challenges) which must be in conditions of be solved or to contribute to solve by the students or graduated of the program, as in the present as in the future (during the next 5-10 years, approximately). As it is described in Tobón (2010 a, 2011a,b,c,d), these problems are needs of formation with respect to the personal development, the family, the community, the society, the organizations, the enterprises, the laboral-professional world , the environment, the recreation, the sport, the culture, the politics, the communications, the technology and the science.

To address. Product: report with the study of the external context. Criteria:

1. The challenges of the external context are identified by considering the different areas in which the discipline or profession intervene.
2. The challenges of the external context are determined as in the present as in the future (next 5-10 years, approximately), based on trustworthy sources.
3. The challenges of the external context are identified by considering the local, national, international area.
4. The challenges of the external context are determined by looking for the improvement and innovation of the program of studies, according to the challenge of forming entrepreneurial people.

To plan. The study of the external context is done under the following key actions (Tobón, 2010 a).

- Action 1. Revision of documents about professional tendencies.
- Action 2. Revision of laboral offers published in different media.
- Action 3. Revision of documents with respect to appropriate competencies for the established program in other projects.
- Action 4. Revision of scientific and/ or academic articles about research tendencies.
- Action 5. To search in a group of experts the problems which must be solved by the students and the graduated ones. These expert people must know the different areas in which the graduated people of a program can develop (Tobón, 2010a) The experts must have, at least, two of the following characteristics (Tobón, 2010a):

1. 8 years of experience in their professional exercise at least, according to the academic program to be structured by competencies.
2. To have worked in at least two areas of the profession to have a more global vision.

3. To have participated in a management process of human talent with professionals of the same area (selection, training, evaluation, compensation, and/or promotion processes).
4. To have participated in some relevant study of the context for the program of studies.

- Action 6. When the information of the experts is not enough or it is necessary to have more details about the occupational and professional tendencies, there will be massive tests to some sample population of employers, teachers, students and graduated. According to CIFE Institute´s experience, this last one is not always necessary to be done, when the former actions are followed.

To act:

1. To deal with the problems which are present during the process with pro activity and with a positive thought.
2. To apply, at least, the related actions with the documental study and the enquiry to experts. This can be step by step or in a parallel form.
3. To apply interviews and surveys in a big scale only if it is necessary, after setting into practice the other actions which imply less resources and human talent.
4. To systematize the challenges which the students and graduated students of the program have to face to, if possible by areas of performance.
5. To determine the general actions which must be implemented in the program, to achieve the boarding of the identified challenges of the context.

To evaluate-to communicate:

1. To verify to have had the necessary sources for the study of the external context according to the established criteria.
2. To verify that the systematized and found information tackles the established criteria.
3. To value if the challenges which were determined correspond to the information obtained from the consulted sources.
4. To value if the proposed actions to deal the challenges with in the program are feasible to be implemented.
5. To document the study of the external context, as a level of a followed process as the achieved results.
6. To socialize the final document of the study of the external context, with the educative community through several means and to look for the community to give reflections and suggestions to improve these aspects.

Process 4. Process of graduation

Concept. All the educative program must determine with clarity the graduation process, with the objective of ensuring the students to have at the end, a solid, ethical project of life and the necessary competencies to face the challenges of diverse contexts in which they must act. This implies, mainly, to have a graduation profile with the competencies to form in the students, to determine the accreditation process of such competencies to ensure the social, and laboral-professional linking and to implement actions of reinforcement if necessary.

The competencies of the graduation profile are the ones which orient the design or reform of the curricular map, the formative projects, the actions of mediation of the teachers and the necessary resources management. The graduation profile then, is the nautical letter which orients to principals, teachers and students about the formative process and the academic management.

The process of graduation is carried out based on the study of the internal and external context done in the previous process. It has as an axis the collaborative work and it looks for key agreements with the academic community to ensure it effective implementation (principals, teachers, and students).

To address. Product: report of the graduation profile by competencies with the actions of reinforcement.

Criteria:

1. The graduation profile describes the competencies which must have the graduated ones of the program.
2. The competencies are described according to a determined methodology.
3. The actions of reinforcement agree with the educative model, the competencies of graduation and the resources of the institution.

To plan:

1. To review the study of the internal and external context.
2. To determine what it is expected from the graduated, according with such study of the context.
3. To identify the essential competencies which the students must show at the moment of finishing the program.
4. To establish actions of reinforcement of the graduation competencies as well as the social and laboral-professional connection.

To act:

1. To analyze how the graduation profiles have been traditionally constructed and to take into account these learning in the new curriculum by competencies.
2. To reflect upon the impact that the graduation profiles have had in the educative programs.
3. To establish the denomination of the program, from the study of the internal and external context and to justify it. The denomination of the program must be based in its contents, level of formation, purposes and profile, I correspondence with the educative model and the legal norms. The name of the program must consider its emphasis or the object of transformation which it is going to deal with. For that, it is necessary to consider the history of the profession, its general areas of performance (command of the competence) and its code of ethics.
4. To elaborate the graduation profile according to the frame 3.6.

Frame 3.6. Graduation profile

Name of the program	
Specific competencies:	
Competence 1.	
Competence 2.	
Competence 3.	
Competence 4.	
Competence 5.	
Competence 6.	
Generic competencies	
Competence 1.	
Competence 2.	
Competence 3.	
Competence 4.	
Competence 5.	
Competence 6.	
Actions of reinforcement of the competencies at the end of the program	*Actions of social and/or laboral-professional connection.*

To evaluate-to communicate:

1. To establish if the graduation profile responds to the challenges and requirements of the context.
2. To verify that the name of the program corresponds to the educative level, with the areas of performance and with the cycle of formation.
3. To document the process of construction and/or improvement of the graduation profile.
4. 4. To socialize the document of the graduation profile and to agree its implementation.

Guidelines to describe the graduation profile

1. Name of the program. The name of the program is set out. It has to fulfill the legal norms, to consider the formative cycle and to have into account its object of action.

2. Writing of the specific competencies:

- They are written with four key elements: verb of performance (if possible, in present tense), conceptual object, finality and condition of the context (see frame 3.7).
- They are determined with base on the study of the internal and external contexts.
- It is recommended not to have many specific competencies (between one and 20 is enough), although the number can vary according to the needs and the program.

When there are specific or generic competencies of a single area, they are organized of Command of competencies, which are global competencies or fields of performance formed by two or more competencies of near nature. The commands of competencies articulate around general problems and related activities with a determined task in the social life, the disciplinary field, the laboral and professional area where reality is continuously asked to articulate from the interrogation of the formative process (Tobón, 2010 a).

3. Writing of the generic competencies:

Frame 3.7 Suggestions to write down a competence.

Verb of performance	*Conceptual object*	*Finality*	*Condition of the context*
It indicates the performance that a person must do.it is recommended to use verbs of performance such as: to manage, to administer, to execute, to measure, to design, to propose, and to do. Verbs such as: to identify, to know, to recognize, to observe, to understand, etc must be avoided.	It indicates objects or situations where the situation appears.	It is to describe the what for of the competence. There can be one or several finalities.	They are the minimum referents of the context which must be achieved in the competence.
Example:			
	Command of the competence: Research		
Verb of performance	*Conceptual object*	*Finality*	*Condition of the context*
Manage	Projects of research	To build knowledge and to solve certain problems of the context	Research method and revision of the antecedent and the bibliography.
Competence: manage projects of research to build knowledge and to solve certain problems of the context, based on a research method and revision of the antecedents and the bibliography.			

Note: the competence of research emerges from the disciplinary and epistemological analysis of the research project.

- They are common and transversal to several areas, disciplines, occupations and professions. They are essential for professional mobility, the life in society, the continuous formation and the management at any field.
- They are written in the same form of the specific competencies.
- They are determined with base on the study of the internal (institutional) and the external contexts. In the studies of CIFE institute, eight key generic competencies have been identified in the formation of the citizens, which are present in the frame 3.8. These competencies have into account the proposed competencies for the Alfa Tuning project and the 6X4 project. In each one of the generic competencies of frame 3.8 ethics is present, which is transversal to all the performance.

1. 4. Actions of reinforcement at the end of the program. The actions to reinforce the specific and generic competencies in the students are described at the end of the program of formation, only if it is necessary.
2. 5. Actions of social and laboral-professional connection. In the frame 3.9 (p. 214) are described the actions to ensure the social and/or laboral-professional connection of the students from their formation.

Frame 3.8 Essential generic competencies proposed by the CIFE Institute.

Number	*Competence*	*Main problems-challenges*	*Description*	*Some criteria*	*Some evidences*
1	Self-management of the formation.	· How to face the continuous challenges of updating , towards the technological , scientific and laboral changes? · Which strategies must be implemented to have an appropriate formation before the changes of the context?	Self- manage the own information along the life to reach a personal chievemt and the established goals by facing the challenges of the context and by considering the opportunities.	· To plan the process of formation, according to the challenges of the context and their projects of life. · To look for necessary resources for their formation, according to the possibilities of the context and their needs. · To do the processes of formation with perseverance until reaching his goals. · To evaluate his formation and to do actions of improvement according to determined goals.	· Document with a plan of formation. · Document with the planning of the ethical project of life. · Evidences of the processes of formation carried out.

Number	Competence	Main problems-challenges	Description	Some criteria	Some evidences
2	Oral and written communication	· How to communicate in the different contexts by using appropriate codes? · Which tools to apply to generate coexistence form the communication?	To use the oral and written language to communicate with understanding in varied social and cultural contexts by using different codes and tools in the frame of a metacognitive process.	· To identify the key ideas in a text or in an oral discourse and to infer conclusions from them. · To write down reports in which to analyze processes and situations of the life, according to syntactic and semantic norms of the language- · To express ideas and concepts of oral form, achieving people to understand a message intended to be transmitted and by considering the communicative requirements of each situation. · To apply several communicative strategies according to the conversational	· Written report about a problem of the context. · Register of an exposition in public. · To summarize and to analyze a discourse. · Register of a socio drama.

Number	Competence	Main problems-challenges	Description	Some criteria	Some evidences
				partner, the context and the objectives which are followed. · To communicate with respect and cordiality with other people, by considering the challenges of the diverse social situations. · To communicate assertively by handling diverse communicative situations which appear, which implies the proactive tackling of conflicts.	
3	Oral and written communication in a second language.	· To do social, research or economic projects, with people who have other languages and cultures? · How to appropriate from information and knowledge in the globalized world?	To use a second language to communicate in oral and written form and to have the possibilities of interacting with other societies, by considering the own criteria of such a	· To identify the key ideas in a text or in an oral discourse in a second language and to infer conclusion from them. · To write down reports in a second language in which to analyze	· Written report about a problem of the context. · Register of an exposition in public. · To summarize and to analyze a discourse. · Register of a socio drama.

Number	*Competence*	*Main problems-challenges*	*Description*	*Some criteria*	*Some evidences*
			language, the ethical commitment and the challenges of each situation and context.	processes and situations of the life, according to the synthetic and semantic norms of such a language. · To express ideas and concepts in oral form in a second language, by achieving people to understand the message to be transmitted and by considering the communicative requirements of each situation. · To apply several communicative strategies in the frame of a second language, according to the conversational partner, the context and the goals.	

Number	Competence	Main problems-challenges	Description	Some criteria	Some evidences
				· To communicate in a second language with respect and cordiality with other people, by considering the challenges of several social situations. · To communicate assertively in a second language , by handling the different communicative situations which may appear, which implies the proactive dealing of conflicts.	

Number	*Competence*	*Main problems-challenges*	*Description*	*Some criteria*	*Some evidences*
4	Collaborative work and leadership	· How to achieve synergy in a team, in the frame of the social and organizational demands? ·Which mechanisms to implement to solve the conflicts which appear in a team and to generate agreements?	To do collaborative activities and to lead projects to reach a determined goal, with planning and well defined objectives, in different contexts and with ethical commitment.	-To conceptualize what team work is, its characteristics and responsibilities, by considering the challenges of the context. - To understand the process of planning of activities in team, according to some methodology. -to participate by doing activities in a determined team, with acceptance of the differences and assertive communication, according to determined objectives. To contribute with the team to have a shared vision and a clear program of work, by participating in the	Conceptual cartography of team work. -Working plan elaborated by the team. -Report of activities elaboratedby the team. -Register of goals reached by the team. -Report of solution of conflicts. -Register of a socio drama.

Number	Competence	Main problems-challenges	Description	Some criteria	Some evidences
				analysis and creative solution of conflicts. -To have a sense of challenge so that the team reaches higher goals, according to the shared vision with the team. - To coordinate processes of planning and execution of activities and projects, according to the contextual challenges to the ethical project of life. -To identify difficulties in the team work and to propose clear and feasible solutions, by assuming responsibilities to overcome such difficulties. -To have ethical commitment in the	

Number	*Competence*	*Main problems-challenges*	*Description*	*Some criteria*	*Some evidences*
				treatment with people according to the universal values. -To have relationship with others though an assertive communication.	
5	Management of the information and of knowledge	· Which tools to use to generate knowledge from information? · How to contextualize the knowledge to solve local problems.?	To process the information related with a determined process of reality, with the objective of generating understanding and knowledge. To have as a referent the challenges of the context, the tools of planning and management and the technologies of information and of communication.	· To identify the requirements of information and of knowledge in the context, according to a determined goal. · To process the information and the knowledge through the use of a computer to a level of user. · To do collaborative activities through the use of internet (e-mail, chat, video chat, web pages, etc) and of land line and mobile line. · To interpret information according	· Report of the processing, interpretation, argument and solution of a problem of the context, by applying the technologies of information and of communication.

Number	*Competence*	*Main problems-challenges*	*Description*	*Some criteria*	*Some evidences*
				to determined conceptual and contextual referents. . To analyze the contextual processes of a systemic form and to consider this when dealing with the activities and the problems. . To process the information with base on a determined methodology, the established goals, the technologies of information and of communication and the ethical commitment. . To argue the form of dealing with activities and problems of the context, by considering determined knowledge.	

Number	*Competence*	*Main problems-challenges*	*Description*	*Some criteria*	*Some evidences*
				· To propose systematic solutions to the problems from the analysis of information and of determined knowledge.	

Number	Competence	Main problems-challenges	Description	Some criteria	Some evidences
6	Entrepreneurship	· How to achieve the projects to keep in the time and to face the difficulties strategically? · How to change the difficulties and menaces in opportunities to do projects of entrepreneurship? · How to achieve the entrepreneurial projects to be based on a, ethical project of life and in the facing of the challenges of the context.	To execute projects to solve problems of the context, with perseverance to reach the proposed goals, by having into consideration the established opportunities.	To identify problems in the context and to change them into opportunities to establish new and appropriate social or economic projects. -To design projects with creativity and innovation, according to the requirements of the context. -To implement the actions of the projects taking into consideration the proposed goals, the defined administrative process and the conditions of the surroundings. -To make the most of the benefit and the projects with perseverance and to face the difficulties	Report with the planning of an entrepreneurial project. -report of execution of an entrepreneurial project. Register of socialization of an entrepreneurial project.

Number	Competence	Main problems-challenges	Description	Some criteria	Some evidences
				strategically. -to perform ethically in the entrepreneurial processes by considering the diverse challenges of the context.	
7	Research	· How to generate knowledge and/ or new methodologies to answer to disciplinary, professional, social, environmental, entrepreneurial or economic problems? · How to work in a collaborative work to have impact on the research?	To solve problems of the context through a determined methodology of research, to generate knowledge and to act with a major impact in the reality, by considering the accumulated knowledge, collaborative work	· To argue the different components of the research concept, with conceptualization, central characteristics, differences, division, connection, methodology, etcetera. · To plan a project of research (or of intervention with a research component) according to a determined problematic and methodology of presentation of projects.	· Report of the conceptual cartography of a research project. · Report of the full planning of a research project. · Report of the e execution of a research project.

Number	*Competence*	*Main problems-challenges*	*Description*	*Some criteria*	*Some evidences*
				· To execute a research project in a pro active and strategic attitude to face the difficulties which emerge in the process. To do opportune adaptations, according to the expected results. · To systematize the given information by the project under a determined method, according to the objectives and methodology of the same project. · To socialize the results of the research, through different strategies (speeches, videos, articles, books, etc.) talking about the problem, the methodology, the	

Number	Competence	Main problems-challenges	Description	Some criteria	Some evidences
				results and the reached conclusions. · To perform ethically in all the stages of the research process, according to the nature of the project, the ethical codes of research and the antropoetics	
8	Management of quality	· How to generate a culture of quality and to keep it through collaborative actions, to achieve the goals, with a a high degree of pertinence, equity, efficacy and efficiency?	To manage the quality of the processes and products of a project to generate a high grade of possible satisfaction and pertinence to an internal and external level, by considering the maximum standards in the context.	· To determine the criteria which are considered in the management of organizational quality, according with the goals of the organization and the current norms. · To auto motivate about quality management, what allows to do activities with dedication, perseverance and	Report of quality assurance in an activity or project, by indicating the fulfillment of criteria or standards and the actions of implemented improvement.

Number	*Competence*	*Main problems-challenges*	*Description*	*Some criteria*	*Some evidences*
				expected quality. · To plan processes of quality management by considering the organizational context and the established criteria. · To execute concrete actions of quality improvement in a determined organizational process, by doing opportune adaptations and by considering the expected results. · To act according to the ethical code in the field of quality management,- To have as a base the metacognitive reflection. · To evaluate the quality management with base on determined criteria.	

PROCESS 5. Process of entry and reinforcement of the entry competencies

Concept. Having as a referent the graduation profile, it is necessary to construct the entry profile. For this, the competencies which the students must have are determined at the beginning of the program of studies, with the objective of achieving the success in their formation. Concretely, in the graduation profile are described the basic, specific and generic competencies that are expected in the students at the moment of entering the program (although it is not always necessary to have the three types).

It is important to keep in mind that the graduation profile is desirable for selection processes, but, overall, it is basic for doing the diagnostic evaluation once the students enter the program. It helps to determine lacks in their formation of the previous cycles, and thus, to offer them the required help.

To address. *Product:* report of the entry profile by competencies with reinforcement actions.

Criteria:

1. The entry profile describes the competencies which the students must have, according to the study of the internal and external contexts.
2. The competencies are described according to a determined methodology.
3. The reinforcement actions match with the educative model, the graduation profile and the resources of the institution.

To plan:

1. To review the study of the internal and external contexts.
2. To determine what is expected from the student so that the graduation profile is achieved.
3. To identify the essential competencies that the students must show at the moment of entering the program.
4. To establish actions of reinforcement of the entry competencies.

Frame 3.9 Example of a graduation profile.

Name of the program: psychology	**Length: 9 semesters**
SPECIFIC COMPETENCES	
Command of the competence 1: complex psychological evaluation	
Synthesis of the problem of the context	*Competences*
· How to evaluate the processes of mental health according to the biological and social contexts, on a base of an integrative model of human health, with contributions of different disciplines? · How to favour in the people the development of metacognitive strategies to generate processes of self regulation, which lead to the self-management of mental health? · Which actions to implement to achieve psychosocial evaluation to be a formative process and to contribute to improve the quality of life in the people?	1.1 to evaluate the implied psychological processes in the biopsicosocial health to promote the quality of life, with base on appropriate instruments, the ethical performance, complex thinking and the contributions of psychology and other disciplines. 1.2 To evaluate the superior psychological functions to determine possible needs of help, with appropriate instruments, assertive communication, and an integrative model of evaluation with the contributions of psychology and of other similar disciplines.
Command of the competence 2. Psychosocial intervention: assessory and treatment	
· How to give attention in mental health to improve the quality of life with the bio psychosocial paradigm? · Which mechanisms to implement so that people strengthen their ethical project of life from the social intervention?	2.1 To offer psychosocial intervention to improve the quality of life, with the psychology and the contributions of other disciplines, with base on the antro poetic performance. 2.2 To give psychosocial attention to people with problems of physical health to contribute to increase their quality of life, with the integral evaluation, the empathy, and the contributions of different disciplines. 2.3. To do a neuropsychological attention, to improve the quality of life, with base on an integral diagnostic of the person by applying the complex thinking. 2.4 to manage actions of prevention and promotion in the field of the bio psychological health, to achieve high levels of quality of life in people, with responsibility, complex thinking and with theoretical and methodological contributions as in psychology as in other disciplines.

Name of the program	**Established Pre-requisites in the law to enter.**		
	Description	*Essential criteria*	*Key evidences*
Minimum basic competencies			
· Basic competence 1			
· Basic competence 2			
Specific competencies			
· Specific competence 1			
· Specific competence 2			
Generic competencies			
· Generic competence 1			
· Generic competence 2			

To act:

1. To elaborate the graduation profile according to the frame 3.10.
2. To implement two types of activities of reinforcement according to their needs:

Frame 3.10. Format of the entry profile and reinforcement of the entry competencies.

Name of the program	**Established Pre-requisites in the law to enter.**		
	Description	*Essential criteria*	*Key evidences*
Minimum basic competencies			
· Basic competence 1			
· Basic competence 2			
Specific competencies			
· Specific competence 1			
· Specific competence 2			
Generic competencies			
· Generic competence 1			
· Generic competence 2			
Evaluation of the entry profile			
Action of reinforcement			

Notes:
1. The basic competencies are: reading, writing and arithmetic
2. The specific competencies are proper of an area, discipline, occupation or profession.
3. The generic competencies are transversal, essential competencies to perform in any field (for example, entrepreneurship).

a. Activities in the formative, mandatory spaces. They can be for all the students or only for those who need them. With these activities, concrete achievements are expected, those which do not require too much time or the continuous help of teachers.
b. Activities in optative formative spaces along with mandatory subjects. These must be only done by the students who present substantial weaknesses in the entry profile.

In both cases, it is recommended to evaluate the activities of reinforcement, and this represents part of the working academic load of the students. Therefore, they are more valued by the students and the evaluation itself contributes to the achievement of goals.

To evaluate- to communicate:

1. To establish if the entry profile responds to the challenges and requirements of the context.
2. To determine if the entry profile has the essential competencies to achieve success in the program.
3. To document the process of construction and/or improvement of the graduation profile.
4. To socialize the document of the entry profile and to agree its implementation (frame 3.11)

Frame 3.11 example of an entry profile and reinforcement of the entry competencies.

Name of the program: Psychology		**Established Pre-requisites in the law to enter:** Previous studies: high school degree	
Minimum basic competencies			
	Description	Essential criteria	Key evidences
Oral and written communication	To use the oral and written language to communicate with understanding in varied social and cultural contexts, by using different codes and tools in the frame of a metacognitive process.	· To identify the key ideas in a text or oral discourse and infers conclusions from these. · To write down reports in which there is an analysis of processes and situations of the life, according to syntactic and semantic norms of the language. · To express ideas and concepts, in oral form, for the people to understand the message they want to transmit with the communicative requirements of each situation.	· Register of an oral report about the analysis of a problem of the context. · Written report about an analysis of a problem of the context.
Command of the competence: Mathematics			
Competence of operative mathematics	To apply the fundamental operations of mathematics to analyze and solve problems of everyday life, in accordance with the disciplinary field.	· To calculate the arithmetic operations by using a standard algorithm. · To argue the solution of the problems of the context through mathematics	· Written report about the analysis of a problem of the context.

Statistics competence	To organize the information of the context through tables of descriptive data, by doing averages, identifying and relating variables, and organizing quantitative data in general, to analyze and to solve problems of everyday life, in accordance to the basic tools of the area.	· To solve problems with base on the descriptive statistics. · To argue the solution of problems based on the descriptive statistics.	· Written report about the analysis of a problem of the context.
Generic competencies			
Management of the information and of the knowledge.	To process the information related to a determined process of reality, with the objective of generating understanding and knowledge, by having as a referent, the challenges of the context, the tools of planning and management and the technologies of information and communication.	· To identify the requirements of information and of knowledge in the context, according to a determined goal. · To process the information and the knowledge based on the use of computer and in a user level. · To interpret the information in accordance with determined conceptual and contextual models. · To analyze the contextual processes in a systemic form. This takes into consideration to deal with activities and problems.	· Register of an oral report about the analysis of a problem of the context. · Written report about the analysis of a problem of the context.

Name of the program: Psychology		**Established Pre-requisites in the law to enter:** Previous studies: high school degree	
	Description	Essential criteria	Key evidences
		· To process the information with based on determined methodology in the established goals, in the technologies of information and n the communication and in the ethical commitment. · To argue the form of dealing with the activities and problems of the context, by considering determined knowledge. · To propose systemic solutions to problems, since the analysis of information and of determined knowledge.	
Evaluation of the entry profile	The entry profile will be evaluated in the formative spaces of the first year, through the described evidences.		
Actions of reinforcement	In the program of psychology there wil be two actions of reinforcement: · Activities in the mandatory subjects of the first year. · Activities in optative subjects for the students who present meaningful weaknesses.		

MACROPROCESS III. MANAGEMENT OF THE FORMATIVE PLAN

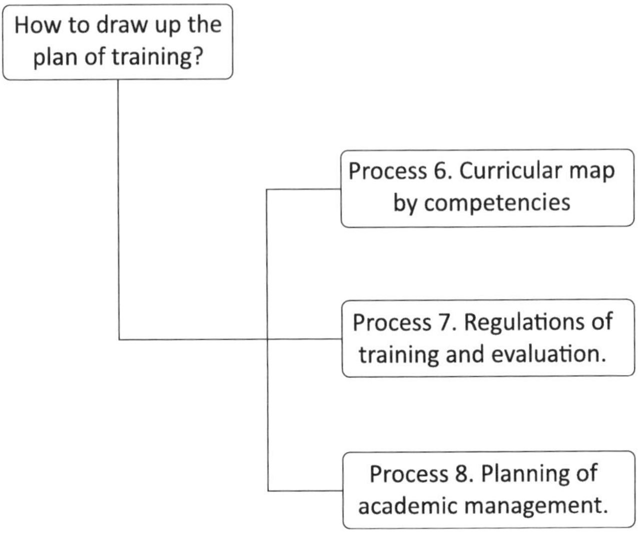

Process 6. Curricular Map by Competencies

Concept. Once the entry and graduation profile exist, the next step is to draw up the curricular map or the training map, also known as curricular net. The Curricular net consists on the graphic representation of the plan of studies, with the commands of competence, the academic periods, the mandatory formative spaces, the optative formative spaces and the length of each formative space, with the objective of achieving as the entry as the graduation profile.

To address. *Product:* report with the curricular map of the program. Criteria:

1. The curricular map of the academic program has in mind the educative model of the institution.
2. The curricular map of the academic program has in mind the study of the internal and external contexts.
3. The curricular map of the academic program considers the entry and graduation profiles.

To plan:

1. To review the chronogram of work regarding the curricular map and to make the possible adjustments in the activities, their

length, the resources and the form of getting to key agreements which assure their effective implementation.
2. To look for basic elements to design the curricular map as the educative model and the graduation and entry profiles.
3. To have clarity in the methodology which is going to be applied to make the curricular map.
4. To determine how to overcome the problems of the traditional nets by subjects in the new curricular map.
5. To establish the length of the program in accordance to the legal norms and institutional policies.

To act:

1. To make the curricular map based on a determined structure, by establishing the necessary formative spaces to assure the achievement of the entry and graduation profile.
2. To establish the necessary formative spaces to form the competencies of the profile, as in a direct as in a transversal form in each academic period.
3. To assign time to each one of the formative spaces of the curricular map, by considering the average time that the students require to achieve the criteria and to do the key activities. For this, it is necessary to draw up the basic aspects of each formative space and to have in mind the own experiences, the experiences of other colleagues and the studies that exist in the field.
4. To determine the logic sequence among the formative spaces looking for the pre-requisites and the co-requisites to be strictly necessary.
5. To establish the requisites of final degree (for example, the evaluation of the competencies).
6. To implement, in the studies plan, innovations which overcome the traditional nets made by subjects. To consider specially concrete actions of flexibility and the work with transversal projects.

The structure which appears in the frame 3.12 is one of the most simple and it is appropriate to begin a program by competencies since the basics.

Command of the competence	Period 1	Period 2	Period 3	Period 4	Requirements for final degree: integral evaluation of the competencies.
Command of the competence 1.	Formative space 1.1		Formative space 1.2	Formative space 1.3	
Time	Time		Time	Time	
Command of the competence 2	Formative space 2.1	Formative space 2.2			
Time	Time	Time			
Command of the competence 3	Formative space 3.1		Formative space 3.2	Formative space 3.3	
Time	Time		Time	Time	
Command of the competence 4	Command of the competence 4	Command of the competence 4			
Time	Time	Time			Time
Integral professional practice					Formative space of integral practice. time
Time			Time	Time	Time
Total time per period	Total time per period	Total time per period	Total time per period	Total time per period	Total time per period

Command of the competence		Semester 1	Semester 2	Semester 3	Semester 4
Complex psychological evaluation. How to evaluate the processes of mental health, according to the biological and social contexts, which have as a base an integrative model of human health with contributions of different disciplines?	Formative projects		To evaluate the process of psychosocial health of a person and to determine an intervention plan.		
	Credits		6	6	6
Psychosocial intervention How to give attention to mental health to improve the quality of life, by having in minds the biopsychosocial paradigm? Which mechanisms to implement so that people strengthen their ethical project of life from the psychosocial intervention.	Formative projects	To interpret and to argue problems related with the psychological functioning.	To interpret and to argue problems of mental health with base on historic contributions of the psychology.	To interpret and to argue problems of mental health based on the evolutive development of the human being.	To give assessor in mental health to one person or group.
	Credits	6	6	6	6
Management of the human talent. How to promote the human talent development from the antropoetic, and social self-management commitment, on a frame of the homeland by considering the unit in the diversity.	Formative projects	To build and to enhance the ethical project of life			
	Credits	5			

Communication and research. How to apply the scientific thought since the communicative process, by developing interpretative, argumentative, and propositive processes	Formative projects	To apply cognitive, metacognitive, and cooperative strategies in reading, oral expression and production of written texts.			
	Credits	5	5	5	5
English with the European Standards. How to live in the knowledge society?	Formative projects	Competence a1	Competence a2	Competence b1	Competence b2
	Credits	6	6	6	6
Human and universal training. How to have a universal, scientific, artistic training and ample culture, to live in the home land and to give to the great problems of humanity?	Credits of free configuration	5	5	5	4
Development and consolidation of professional competences. How to develop the professional competences of an integral form?	Formative projects				
	Credits				
Total credits per period. *Total of the carrier:* 240 credits		27	28	28	27

Management of human talent How to promote the development of human talent since the antropoetic commitment and the social self management, in the frame of the homeland, by considering the unity in the diversity?	Formative projects	To implement the management of human talent in educative institutions	To manage the human talent in processes of social development			To implement the management of human talent in the organizations. This module is shared with the masters of management of human talent
	Credits	6	6	6	6	6
Communication and research How to apply the scientific thought since the communicative process, by developing interpretative, argumentative and propositive processes?	Formative projects		To design, execute, and publish a project of research.			
	Credits		5	5	5	
English based on the European standards. How to live in the knowledge society?	Formative projects					
	Credits					

Management of human talent How to promote the development of human talent since the antropoetic commitment and the social self management, in the frame of the homeland, by considering the unity in the diversity?	Formative projects	To implement the management of human talent in educative institutions	To manage the human talent in processes of social development			To implement the management of human talent in the organizations. This module is shared with the masters of management of human talent
	Credits	6	6	6	6	6
Communication and research How to apply the scientific thought since the communicative process, by developing interpretative, argumentative and propositive processes?	Formative projects		To design, execute, and publish a project of research.			
	Credits		5	5	5	
English based on the European standards. How to live in the knowledge society?	Formative projects					
	Credits					

COMMAND OF THE COMPETENCE		Semester 5	Semester 6	Semeste 7	Semester 8	Semester 9, Professional degree
Human and universalist training. How to have a universalist, scientific, artistic, and ample culture training to live in the homeland and to give to the big problems of humanity?	Credits of free configuration	4				To elect between these two options: -Project of reinforcement of the graduation competencies. -Project of the masters management of human talent.
Development and consolidation of the professional competences. How to develop the professional competencies of an integral form?	Formative projects					
	Credits			7	7	7
Total credits per period. *Total in the carrier:* 240 credits.		25	28	28	28	21

Note. The program has 240 credits in accordance with the system of credits of the European Space of Superior Education.

Next, we will explain briefly the components of the curricular map.

1. Commands of the competence. All the curricular map is oriented to do reality as the graduation competencies as the entry competencies. This can be done in an implicit or explicit manner.

In the case of the curricular map here presented in the frame 3.13, the commands of the competence are included implicitly, which are areas of performance prioritized in the training. They are based on a macro problem and are composed of alike competencies, which are formed along the plan of studies.

2. Academic periods. They are the spaces of time in which the training is divides in a determined educative cycle. They can be a two month period, a trimester, semester or years, etc. in the case of the plan of studies presented in the frame 3.13, the academic periods are semesters, which is the most recommended one.

3. Training spaces. They are the minimum structure in which the curricular map is organized. This can be by subjects, modules, learning based in problems or formative projects. Moreover, since a flexible point of view, there can be a combination of methodologies of curricular structure.

There are several possibilities of formation of the competencies in the training spaces (Tobón, 2010 a):

Model 1. One Competence, one training space. For each specific competence of the graduation profile, there must be established a training space. Then, there are going to be as many training spaces as competencies of the graduation profile.

Advantages of this model:

- There is clarity about the competencies to be developed in the training spaces.
- There are major possibilities to develop the competencies because each training space has a competence as a goal.

Model 2. One competence, several training spaces. Several formative spaces develop a specific competence of the graduation profile. This can be done according with the following options:

- Option 1. The criteria are distributed in each one of the training spaces. In this form, as long as these criteria are formed, the established competencies are developed.
- Option 2. In each one of the training spaces is distributed a determined level of performance of the competence to be developed. For example, in a space is formed until the resolutive level, in another, until the autonomous level, etc.

Model 3. Several competencies, a training space. A training space can develop several specific competencies of the graduation profile.

Advantages of this model:

- It can be possible to have few training spaces and to do a more integrative training.
- In a training space, the students learn how to be competent in several areas.

Model 4. Several competencies, several training spaces. Several training spaces contribute to develop several specific competencies of the graduation profile and criteria are taken into account. Therefore, the criteria are the ones which help to determine the specificity of each training space in the curricular map.

Advantages of this model:

- High degree of difficulty in the setting of curricular map.
- Possibility of continuity and that in the following one, some pendant criteria from a previous training space can be accredited.

Independently of these options, in each training space there is contribution to develop one or several competencies. The most recommendable is to have one or two generic competencies by space in order not to complicate the process a lot.

In the example of the frame 3.13 the models 1 and 2 are tackled, with base in the methodology of training projects. If in the educative institution the teachers and the principals have a high degree of formation in didactics and they work in collaborative form, there is possibility to advance towards major levels of curricular integration and to work with the models 3 and 4. For example, training projects can be established to form two or more specific competencies.. The practice has taught us that it is better to go step by step and to work with projects, highly integrators, only when there is experience and formation in the field. From the opposite view, it is possible to go directly to failure.

The Training projects (PF) are articulated sets of activities to identify, to interpret, to argue and to solve the problems of the context, with base on planning, execution and socialization of one or several relevant products through the articulation of to know how to be, how to know, how to do and how to coexist (Tobón, 2010b).

4. Problematic nodes. This component consists on establishing one or several general macro problems of the social, disciplinary, laboral-professional and researching context in each domain of competence, with the objective of orienting the process of training. The macroproblems can be reviewed and adjusted according to the challenges of the context to continue being valid. They can synthetize to be better visualized in the curricular map.

5. Credits. Credits account for the time which is required by an average student to do the necessary activities in a determined training space and to achieve the expected goals., by having in mind the learning time with the teacher and the learning, independent time (without a direct help of the teacher). In this way, there is a change of paradigm in the planning , because it dos not appeal to how much time a teacher requires to teach, but how long an average student requires to develop the competencies. This implies the attendance to classes, but also the independent study outside the classroom, such as library consults, individual and collective work, the preparation o evidences of evaluation, etc.

Therefore, to determine the credits, it is precise to analyze the competence to develop and the proposed activities in each formative space , by having in mind, the experiences of the area and the existent studies about. It is recommended to work with the European Methodology of Credits (ECTS) (European Credit Transfer and accumulation system, European System of credit transfer), in which one credit is equivalent , in average, to 25-30 hours of work, and the learning time is included with the teacher (TAD) and the Independent learning time (TAI).

The relationship between TAD and TAI varies, in accordance to the nature of the competence and the activities of the training space. Normally, it is estimated that per one hour of learning with the teacher, there are two hours of autonomous learning in programs of pre grade. However, this can change in training spaces which require of major help of a teacher, such as training spaces based on the work in the laboratory. This must be done to avoid that the weekly working load of work of the students overpass the rank from 45-50 hours (working load of a student in full time) (Tobón and Garcia Fraile, 2006; Zabala, 2003).

In the curricular map of the frame 3.13 are described the credits in each one of the PF and in each period, according to the approach of credits ECTS. It is important that in Latin America there is an advance in a common system of credits (if possible, to follow the ECTS model, as it is the case of Chile). This would ease the processes of mobility of students and professionists.

6. Requirements for obtaining the degree. In the socioformative approach, it is recommendable that, in the degree requirements, exist an integral valuation of the competencies of the students, according to the graduation profile and that such evaluation is considered into account for the graduation. For example, there could be a policy of giving the degree only to those who show the essential competencies in the resolutive level.

To evaluate-To communicate

1. To determine if the curricular map corresponds to the educative model, the study of the internal and external contexts and the entry and graduation profile.

2. To verify that there are necessary training spaces to develop the expected competencies of the graduation profile.
3. To determine if the assigned time to the training spaces correspond to the required time by the students to achieve the criteria and to learn the expected competencies.
4. To verify that the requirements for the degree correspond with what is required to assure the quality of formation, with the institutional policies and with the legal norms that exist.
5. To document the process of construction and/or improvement of the curricular map.
6. To socialize the document of the curricular map with the educative community and to achieve basic agreements with respect to its implementation.

Process 7. Regulations of formation and evaluation

Concept. The process consists on building and/ or on improving the essential norms which must be fulfilled in the training process and of evaluation of the competencies in the students. These norms are determined by having as a base the institutional educative model, the study of the internal and external contexts , the graduation profile, the entry profile and the curricular map.

To address. Product: Regulation of training an evaluation.
Criteria:

1. The regulations of training and evaluation meets the legal norms established in the area.
2. The regulation takes into account the contributions of the different actors of the educative community.
3. The regulation considers the educative model of the institution.
4. The regulation takes into account the approach of the competencies.

To plan:

1. To review and to adjust the timetable of creation or improvement of the regulation of training and evaluation.
2. To determine the key structure of the regulation of training and evaluation of the students, by considering the legal norms.
3. To limit the process of participation of the educative community in the elaboration and agreement of the regulation of training and evaluation.
4. To establish the necessary actions to fulfill the regulation of training and evaluation.

To act. To do the regulation of training and evaluation by determining, as minimum, the following essential components:

1. Key norms of entry, permanence and graduation.
2. Key norms with respect to the recognition of previous studies.
3. Key norms around the formation of the competencies.
4. Key norms around the process of evaluation of the competencies.
5. Key norms around the practices.
6. Key norms around scholarships, acknowledgements and stimuli to academic labour.

To evaluate-to communicate:

1. To establish if the training and evaluating regulation has into account the approach by competencies.
2. To determine if the regulation corresponds to the legal norms and the educative model.
3. To verify that the regulation has into account the participation of the academic community.
4. To document the process of construction and/or the improvement of the regulation of training and evaluation.
5. To socialize the regulation of training and evaluation with the educative community and to achieve the basic agreements around its implementation.

Process 8. Planning of the academic management

Concept. The Academic Management refers to the necessary policies to implement the curricular map and to do a reality, the institutional educative model. This implies to have clear guidelines around the administrative management, the assurance of quality, the teaching management, research, extension, the work with the graduated ones and the welfare process.

To address: Product: documents with the guidelines and norms of academic management.

Criteria:

1. The academic management fulfills the legal norms established in the area.
2. The academic management has into account the contributions of the different actors of the educative community.
3. The academic management considers the educative model of the institution.
4. The academic management takes into account the approach of competencies.

To plan:

1. To review and adjust the timetable of creation or improvement of the guidelines of academic management.
2. To determine the key structure of the different components of the academic management.
3. To determine the process of participation of the educative community in the elaboration and agreement of the policies of academic management.
4. To establish the necessary actions to achieve the academic management.

To act. To elaborate the academic management with the following components.

- Administrative management
- Management for the assurance of the educative quality.
- Teaching management.
- Management of the research
- Management of extension (work with the community).
- Management of the process of students´welfare.
- Management of the graduated ones (see frame 3.14).

To evaluate-to communicate:

1. To establish if the policies of the academic management consider the approach by competencies.
2. To determine if the policies of academic management correspond to the legal norms and the educative model.
3. To verify that the policies of academic management consider the participation of the educative community.
4. To document the process of construction and/or improvement of the policies of academic management.
5. To socialize the policies of the academic management and to achieve the basic agreements around their implementation.

Frame 3.14 Components of the academic management.

Area of the management	*Area of the management*
Administrative management	· Admission process · Enrollment process · Register of evaluation process · Practice of register process · Graduation process
Management for the assurance of the educative quality	· Self evaluation process · Institutional and programs accreditation process · Certification of quality process
Teaching management	· Collaborative work among professors. · Meetings among professors at the beginning, during, and at the end of the academic period. · Projects of teaching entrepreneurship
Management of the research	· Research lines and responsible ones · Participation of the students in research lines. · Types of research · Agreements with other organizations · Regulation of the patents · Regulation of the publications · Minimum structure of the research projects · Financing of the research projects
Management of extension (work with the community)	· Priority lines to do the projects of work in society. · Minimum structure of the work projects with the society. · Financing of social projects · Participation of principals, teachers and students in projects of work with the society.
Management of the process of students´welfare	· Welfare services for principals, teachers and students. · Evaluation of the welfare services. · Participation of the educative community in the welfare services.
Management with the graduated ones	· Participation of the graduated ones in projects of research and extension · Follow up of actions with the graduated ones. · Participation of the graduated ones in the management of academic quality of the programs · Actions of help to the graduated ones for their social, academic, laboral and professional connection.

MACROPROCESS IV: MANAGEMENT OF THE CURRICULUM IMPLEMENTATION

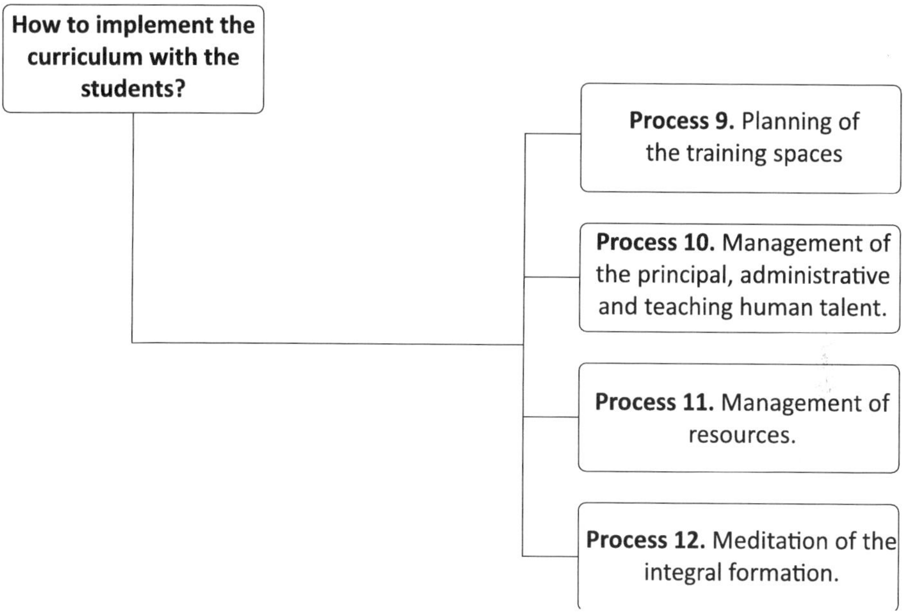

Process 9. Planning the training spaces

Concept. this process consists on planning each one of the training spaces established in the curricular map. To achieve this, one must consider the educative model, the study of the internal and external contexts and the entry and graduation profiles.

A training space consists in a set of learning, teaching and evaluation activities which are done with the objective of the students to develop the expected competencies , with a definite time, and through the mediation of a teacher and the use of determined resources.

The planning of training spaces can be by subjects, modules, learning based in problems, or by formative projects; everything depends on the educative model and on the curricular map which was agreed to be followed (see frame 3.15). Also, there can be a combination of methodologies, on condition that this is done with articulation and complementarity (other perspectives can be found in Pimienta, 2012a, Tobón et al., Didactic Sequences..., 2010 and Zabalza, 2003).

Frame 3.15 Main differences among subjects, modules, learning based in problems and formative projects.

Aspect	*Subjects*	*Modules*	*Learning based in problems*	*Formative Projects (PF)*
Purposes	To learn contents form objectives.	To form competencies from units of learning.	To form the competencies through the interpretation and argument of one or several problems.	To form, at least, one competence through the solution of a problem of the context, in a frame of social commitment, antropoetics, transversality and entrepreneurship.
Structure	They are structured based on themes and subthemes. There is a great emphasis on contents.	The key axis of modules are the learning units articulated among them, in which there are aspects or elements of competencies tackled.	It follows the phases of learning based in problems, such as: ·Analysis of the context Interpretation of the problem. ·Establishment of options to solve the problem. ·Selection and argumentation of the most pertinent option.	The phases of a project are followed: ·Addressing ·Planning ·Performance ·Communication This can be in an implicit or an explicit form. The methodology of the learning based in problems is taken into account, but with an emphasis to solve the problems in the context.

Evaluation	It is based in objectives and contents. Tests and written texts are privileged.	It is done through evidences: · Of knowledge · Of doing · Of product	It is done from the identification, interpretation and argumentation of contextualized problems, and with sense for the students.	It done as the project is executed, with base in the same evidences which it goes facilitating. The evidences are valued with base in learning maps, through the self valuation, covaluation and heterovaluation. This emphasizes in dealing with the process of evaluation from the metacognition made by the student.
Contents	Emphasis in thematic contents. There is a tendency to separate the theory and the practice.	They deal with contents since pertinent situations, by considering the competencies.	The contents are tackled from the interpretation and argumentation of the problems.	The contents are tackled from the addressing, planning, execution and socialization of the project. It has into account the competences to be formed. The students are taught cognitive, metacognitive and sensitization strategies (Tobón, 2010 a) with the objective of achieving success in the solution of problems.

Addressing: product: Document with the essential structure of each one of the defined training spaces in the curricular map.

Criteria:

1. The planning of training spaces agree with the educative model, the study of the internal and external contexts, the entry and graduation profiles, the curricular map, the regulation of formation and evaluation, and the policies of academic management.
2. The planning of the training spaces corresponds to a determined approach or theoric-methodological model of competencies.
3. The planning of training spaces agrees with the modality of education of the program: in person, semi-in face to face, distance or on line.

To plan:

1. To review and to adjust the timetable of planning of the training spaces.
2. To identify the key structure of the training spaces according to a determined methodology of competencies to follow.
3. To determine the training actions for teachers, to plan the training spaces.
4. To establish the process of evaluation and improvement of planning of the training spaces.

Since the socioformative perspective, it is accepted that planning of the training spaces can be by subjects, but it is recommended to transcend these and to advance in more integrative methodologies and of transversal work, such as formative projects.

A formative project consists on making a project with the students in order to make them learn a competence through analysis, understanding and solution of a problem, by looking for a meaningful product which is relevant to the context. The formative projects replace traditional subjects and look for transversality.

It is not necessary to have planned all the training spaces since the beginning, they can be determined before all, the key components. As long as there is work with the students, the most concrete aspects are planned; this makes possible that education is more pertinent and corresponds to the learning needs, by considering the graduation profile.

To act. The structure of the training spaces must have, at least, the following key elements (frame 3.16)

- Title
- Time or credits number
- Competencies to form (to consider at least one generic competence)
- Previous, required competencies.
- Problem or problems to solve.
- Some criteria and evidences.

Frame 3.16 Explanation of the key elements of a training space.

Element	*Explanation*
Title	*It must be related to the competence to form.*
Time	*The number of hours is indicated (it can also be with credits)*
Competencies to form	*The competence or competencies of the graduation profile are pointed out, this contributes to develop a formative space. A generic competence is suggested at least.*
Previous required competencies	*The previous competence or competencies which the students must meet in order to course the training space are described*
Problem to solve	*The problem or problems to solve in the training space are described. To consider the study of the context and the curricular map. This description can be very specific or general.*
Some criteria and evidences	*The essential criteria and evidences of the specific and generic competence or specific competencies, which must meet in the training space, are described.*

In the frame 3.17 is suggested a format to do the planning of a formative project. It can be adapted to the requirements of each educative institution or organization.

Frame 3.17 Format for the planning of a formative project.

Title:		**Time:**	
Competencies to form:		**Previous required competencies:**	
Problems of the context to solve:		**Criteria:**	
Time:	Activities and knowledge	Activities and knowledge	Resources

To evaluate, to communicate:

1. To establish if the training spaces have into account a determined approach of competencies.
2. To determine if the planning of the training spaces correspond to the legal norms, the educative model, the study of the context, the entry and graduation profile, the regulation of training and evaluation, the curricular map and the policies of academic management.
3. To verify the training spaces to be articulated among themselves and them to have a sequence.
4. To document the construction process and/or improvement of the training spaces.
5. To co evaluate the training spaces between teaching pairs and to achieve the continuous improvement.
6. To socialize the training spaces before the educative community and to agree its implementation.

Next, there is an analysis of some characteristics of the formative projects:

1. *Importance of the formative projects.* The formative projects are the methodology that better responds to the nowadays challenges of integral formation, because they make the tackling of different axis of the competencies possible, as in:

- To identify, to interpret, to argue, and to solve problems of the context.
- To learn by doing and by thinking about the meaningful challenges and with a sense.
- To tackle the transversality among areas and disciplines.
- To do collaborative work among pairs, teachers and the community.
- To achieve the inclusive learning of students with different social, cognitive, psychical, activities, etc.
- Besides, the formative projects is the methodology which has more impact in the development of talent of the students, because it allows them:
- To identify with major clarity its areas of interests.
- To learn the necessary elements to achieve success in the life.
- To develop their potential around challenging situations in a real context.
- To brace its self-efficacy, it means, the trust that they are going to achieve their goals, with base on their capacities,knowledge and attitudes.

2. *The formative projects are tackled in all the areas.* As it is well expressed in other texts, (for example, Tobón, 2010a), the formative projects can be worked in all the areas, not only in the practical areas. Whenever there are competencies tackled, the challenge is to solve problems of the context and that makes part of the nature, of the different disciplinary fields. It is sometimes expressed that it is not possible to do projects (for example in mathematics, philosophy, or history) but it is all due, overall, the academic traditional paradigm, because in the nature of these three disciplines is to do applications in projects.

The contextualized problems must be worked in all the disciplines to develop competencies, as in the most abstract mathematics. It is always possible to deal with the challenging of passing from one given situation to an ideal or expected situation, whose solution is not mechanical, but it implies to set into action to diverse abilities, attitudes and knowledge.

Although it is sometimes difficult to find problems of the context, this can be achieved in all the disciplines, with base on creativity, the enquiry, the search of new perspectives and the collaborative work. Here, it is convenient to establish that a problem of the context can refer to:

a. To solve a need.
b. To look for knowledge around a fact.
c. To solve a conflict among approaches and methodologies.
d. To elaborate an object or procedure.

e. To improve a process, object or procedure.
f. To create and object or procedure.
g. To innovate an object or procedure.

A problem of the context also can be of theoretical or symbolic nature. It does not necessarily have to be of the familiar, social or entrepreneurial context. A project can be focused on a problem of this nature, with the same key actions which have been written; addressing, planning, performance, or socialization.

3. *Development and mobilization of knowledge in an articulated form, around contextualized problems.* In the formative projects, the students must learn the necessary knowledge to solve the problems; to know how to be, to know how to do, to know how to know and to know how to coexist, independently of the followed methodology (subjects, modules, projects, etc.).In this sense, the formative projects are not only to do activities, as it is sometimes thought. They also imply that the students build deep concepts and that they tackle key contents of the competencies.

4. *The formative projects are also a didactic strategy.* The micro curriculum can be structured in many forms (by subjects, modules, etc). However, at any methodology, with the students the formative projects can be worked to make possible the development of the expected competencies. This can be done in a single session, in several sessions or in all the training space. Therefore, the curricular map is not necessary to be established by formative projects to tackle this methodology.

5. *The formative projects articulate several strategies***.** Besides being a didactic strategy, the formative projects are a general method which articulates several strategies around the contextualized problems, such as:

- The heuristic Vve.
- The mental maps
- The conceptual cartography
- The socio-drama
- The collaborative work.

Process 10. Management of the administrative and teaching human talent

Concept. The management of human talent is a process to assure people to have a solid, ethical project of life and the necessary competencies to face the challenges of the context, by achieving the personal goals and by contributing to the organizational, social, technologic and environmental development.

In the curricular planning, there must be an implementation of strategies directed to the educative institution, so that it has a principal, administrative, and teaching human talent, with a high degree of suitability, continuous and ethical improvement, which assure the addressing, planning, performance and evaluation-communication of the academic programs, all this based on the institutional, educative model, the studies of the context, and the graduation and entry profiles.

The management of human talent is done by a team. It can be done in the institutional area or in a level of program as such, according to the organizational addressing they have. It is recommendable to have an area of human talent management as in an institutional scale as in each academic unit, because this is a very important quality criterion in the models of management of organizational excellence (for example, the EFQM model, or the Latin American model of quality management). Besides, the accreditation models of educative instittions and or academic programs in this part a lot.

The area of human talent management is the one in charged of establishing the concrete policies and the specific processes to ensure the educative institution to have the suitable human talent which this subject requires. For this, the competencies of the teachers, of the administrative personnel and the principals, as well as the profiles of the positions in the jobs, must be identified. Besides, the processes of selection, evaluation, formation, promotion, pay back, and quality of life must be implemented, by having as a reference the approach by competencies.

To Address. Product: document with the profile of the competencies of each one of the job positions of the institution, and the selection actions, evaluation, training, promotion, pay back, and life quality in the personnel.
Criteria:

1. The management of human talent agrees with the educative model, the study of the internal and external contexts, the entry and graduation profiles, the curricular map, the regulation of formation and evaluation, and the policies of academic management.
2. The profiles of the teachers, principals and administrative personnel are determined with the existent legal norms and the educative model.
3. The concrete actions to carry on selection, evaluation, formation, promotion, paying back, and quality of life of the teachers, principals and administrative personnel, agree with the educative model, the legal norms and the challenges of the context.

To plan:

1. To identify the achievements and aspects to improve the way how the profiles of the teachers, principals, and administrative personnel

are being established in the educative institution.
2. To determine which schemes, obstacles or negative, mental models exist with respect to the management of the human talent of the teachers, principals, and administrative personnel in the educative institution and how they could be modified.
3. To plan how to review the profiles of the teachers, principals and administrative personnel, to structure them by competencies.
4. To establish the processes of selection, evaluation, formation, promotion, paying back, and quality of life of the teachers, principals, and administrative personnel with base on the competencies.
5. To plan how are going to be socialized the strategies of human talent with the educative community and how are going to be discussed or approved by the maximum, principal organism of the institution.

To act:

1. *Position profiles based on the competencies:* This process consists on building the profiles of the positions in the job of the teachers, principals and administrative personnel, based on the generic and specific competencies which they must have for that position, as well as the requirements they must fulfill (such as previous studies and experience).

The profiles established, as that, are the base to carry on the processes of selection, hiring, formation, promotion and remuneration.

To build the profiles of the positions, the following concrete actions are suggested.

a. To determine the necessary competencies that a teacher of any educative institution must have, according to the educative model and the policies established by the maximum , principal organism of the institution. To achieve this, the teaching competencies established by other organizations and which might be adapted to the institution are searched. Some proposals of teaching competencies are general, such as Perrenoud´s (2004a, b); others, on the other hand, are more concrete, such as the one presented in the frame 3.18.
b. To determine the competencies that every principal of the educative institution must posses, in keeping with educative model, and the policies established by the maximum directive organism of the institution. To achieve this, the teaching competencies established in the country or in other countries, or there is a creation of new ones, if necessary (see frame 3.19)
c. To establish the competencies that all the administrative personnel of the institution must meet: auxiliaries, guards, drivers, aids, administrators, etcetera.

Frame 3.18 Minimum competencies which the teachers must meet, according to CIFE Institute

Minimum Teaching Competences	*Minimum Criteria*	*Minimum Evidences*
1. Collaborative work. To do projects or collaborative activities to reach determined goals, in keeping with the educative model, and the action plans of the academic programs	*a)* To have friendly and respectful relationship with the other people, which makes possible to face challenges and difficulties which may appear. *b)* To argue the different methodologies of collaborative work, in keeping with the established objectives. *c)* To form teaching, researching and extension teams, according to the institutional policies. *d)* To do action plans in team, according to the challenges and requirements of the context. *e)* To promote activities of collaborative work in the students, with determined action lines and goals. *f)* To participate in collaborative processes with colleagues, principals, students, according to determined established goals.	*a)* Report of achievements of, at least, three teams of students, with respect to one or several goals and the established action plan. To argue how the teamwork helped to the achievement of goals. *b)* Register about the tackling of a conflict in a team of students and the way it was overcome. *c)* Report of collaborative work with other colleagues with respect to one or several goals and an established action plan during an academic period. To argue how the teamwork helped to the achievement of goals.
2. Communication. To communicate in oral, written and assertive form, with the community, colleagues and students, to mediate in a meaningful form the human, integral formation and to promote cooperation according to the requirements of the educative situations and the institutional functioning.	*a)* To interpret written texts according to the challenges and requirements of a situation. *b)* To produce written texts to communicate determined, suitable messages, with cohesion, and according to the grammar rule norms. *c)* To communicate in an oral form with other people, by promoting understanding, comprehension, and effective activities. *d)* To express clearly, with education and with respect the rights, the emotions, and opinions of others, according to the challenges of each communicative situation *e)* Self reflection about communication and correction of errors, by considering the communicative situation.	*a)* Audio and video of class or speech must be accompanied of a report of evaluation of dissertation by the attendants. *b)* A relevant publication about the experience area during the last two years. It can be an article, a manual, a book or a speech. There must be a follow up of determined norms of styles in the written text. *c)* Report of evaluation of assertive communication, doe by the students of two different groups, as well as by two colleagues at least. This report must indicate as minimum: grade of the school, clarity on the messages, handling of norms, cordiality in the communication, and respect to opinions and rights toward the other who speaks. *d)* Report of how there has been a contribution by the professor, to improve the communication of at least two students or people, by showing the evidences of learning.

cooperation according to the requirements of the educative situations and the institutional functioning.	*d)* To express clearly, with education and with respect the rights, the emotions, and opinions of others, according to the challenges of each communicative situation *e)* Self reflection about communication and correction of errors, by considering the communicative situation.	*c)* Report of evaluation of assertive communication, doe by the students of two different groups, as well as by two colleagues at least. This report must indicate as minimum: grade of the school, clarity on the messages, handling of norms, cordiality in the communication, and respect to opinions and rights toward the other who speaks. *d)* Report of how there has been a contribution by the professor, to improve the communication of at least two students or people, by showing the evidences of learning.
3. Mediation of the formation. To intervene in the processes of formation, teaching and evaluation, so that the students possess a solid, ethical project of life and the develop the necessary competencies to face the challenges of the context, according to the established curriculum.	*a)* To plan the formative processes with a base of diversity in the rhythms and learning styles, as the different sociocultural contexts. *b)* To establish an emphatic relationship and of collaborative work with the students, strengthening the working environment in the classroom. *c)* To do formative sessions with didactic, suitable strategies to the competencies to form, the set criteria the evidences to present and the expected level of performance. *d)* To give special aid to the students who present slow learning or who have difficulties to learn, by considering the educative model of the institution, the expected goals and the process of inclusion. *e)* To teach learning strategies to the students, according to the challenges of the context.	*a)* Report of the planning and execution of a training space by competencies. It must have the tackled competencies, the problem or problems of the context which were worked, the done activities, the evidences of given evaluation by the students and the used resources of effective manner. *b)* Video of at least a complete session under competencies There must be annexed: evidences of the impact of the session-class, report of evaluation of the students, report of the evaluation of two colleagues, and report of the chief of the dependence. *c)* Complete report of the achievement of the competences by two different groups of students, during the academic period, at least the impact of teaching mediation in this. Report of the aid given to a student with difficulties and the reached impact.

Minimum Teaching Competencies	*Minimum Criteria*	*Minimum Evidences*
	f) To implement actions of continuous improvement in the teaching practices, based on the done evaluations and in the innovations in pedagogy or in the educative policies.	
4. Evaluation of the competencies. To value the competencies of the students, to assure the integral formation and a determined profile of graduation, with responsibility, assertive communication, and determined pedagogical and methodological referents.	*a)* To plan the evaluation according to the expected competencies. *b)* To elaborate and /or to adapt instruments of evaluation based on the evidences that the students will present and in determined pedagogical and methodological referents. *c)* To determine the achievement, aspects to improve and level of performance of the competence in the students, according to the application of determined instruments of evaluation. *d)* To give feedback in a respectful and propositive form to students, by considering the competences to form. *e)* To achieve the continuous improvement of the students according to the competencies to form. *f)* To apply the evaluation of the competencies by considering the diversity in learning and in the sociocultural contexts.	*a)* Document with the evaluation plan in a formative space. *b)* Two instruments of evaluation applied to two different groups of students, with evaluation of suitability by these ones and by two colleagues. *b)* At least, three reports of evaluation of competences in the students, by showing their improvement. *c)* Survey to the students about the teachers´ feedback. *d)* Survey to the students about the teachers´ aid to their formation. *e)* Report of evaluation by competencies by considering the diversity.

5. Management of resources and TIC. To manage educative resources and technologies of information and communication to mediate the formation of the students, according to the expected goals and the action plans.	*a)* To assure the spaces of training to be comfortable and organized, and to have the necessary materials and equipment, according to learning goals and by having into account the institutional possibilities and of the context. *b)* To elaborate and /or to adapt the educative materials according to the competencies to form, and by having as a base collaborative work with colleagues and students. *c)* To give the educative materials on time based on the agreed timetable, and in determined quality criteria. *d)* To assume the use of technologies of the information and communication with a critical sense by considering the goals pf integral formation. *e)* To elaborate learning plans with technologies of information and communication, in correspondence to the goals of the plan of studies and the competences intended to form. *f)* To look for, to design and/or to adapt the requires technologies of information and communication for the teaching labor, according to the institutional possibilities, the competences to form in the students and the ethical commitment. *g)* To mediate the learning activities in the students with the use of the technologies of information and communication, according to the competences intended to form and the ethical commitment.	*a)* Two educative materials made for one subjects, module or formative project, to add report with the pedagogical back up of the materials and evaluation by the students or at least by two colleagues. *b)* At least an educative material produced in other context and adapted to the subject, module or formative project, with its corresponding report of pedagogical back up and the evaluation of two colleagues as minimum and of a group of students. *b)* Video of a session or shop done by videoconference, applying the diverse technological tools and following a determined didactic sequence. *c)* To add the repot of evaluation by the students and at least by two colleagues. *d)* Video to show the use of the different tools of an asynchronic learning platform (for example, Moodle, Dokeos, etc). *e)* Report with at least three evidences of evaluation in the students, by applying the technologies of information and communication.

Essential competences of an educative principal.	*Some fundamental criteria.*	*Some key evidences.*
1. Curriculum management. To direct the curricular management to fulfill with the vision and mission of the institution as well as the educative, current policies, according to the definite roles, the current normativity and the needs of the socioeconomic and environmental environment.	*a)* To achieve the participation of all the actors in curricular management (teachers, students, graduated, parents of family, experts, etc). according to the type of institution and the identified challenges. *b)* To address the study of the internal and external contexts, as in the present as in the future, by using an appropriate methodology to the formation of competences. *c)* To direct the construction and/or to improve the educative model of the institution, according to the current, educative policies and the development of pedagogy. *d)* To direct the construction, contextualization, and/or improvement of the plans of study, based on the institutional, educative model and the current educative policies or tendencies in the area. *e)* To direct the continuous evaluation of the curriculum, to introduce improvements and innovations, by considering the results of the evaluating process. *f)* To assume the educative change with flexibility and openness, according to the challenges of the context: disciplinary, social, economical, and political. *g)* To have ethical commitment so that the institutional, educative project answers to the social, pedagogical, academic, political and nearby challenges	*a)* Report of the process of curricular management by implementing, with the corresponding planning, execution and evaluation of results. *b)* Report of evaluations of plans and programs of study, with the corresponding implementation of the improvements.

2. Leadership of the process of formation. To lead the educative process to reach the established goals in the plan of studies, having as a base the indicator of management, the assigned competencies, and the current normativity.	*a)* To direct the planning and set on action of the learning activities and of evaluation, according to the plans of teaching action. *b)* To direct the educative processes based on indicators of management. *c)* To establish alliances with different statements in accordance with the institutional strategy. *d)* To advice to the members of the educative institutions in the application and evaluation of the plans of formation, according to the established goals. *e)* To achieve that the formative plans get transformed in concrete actions with the students, by considering the learning goals. *f)* To have perseverance in the tackling of academic processes and the alliances, until reaching the proposed goals.	*a)* Report about the achievement of educative goals in the institution. *b)* Report of the planned, executed, and evaluated educative projects, with their corresponding impact and relationship with curricular management. *c)* Certificate of at least two established alliances. *d)* Register of assessories to the teachers and visits to the classroom. *e)* Report of the follow up to the academic performance of the students. *f)* Report of an applied survey to the members of the institution around the principal´s role.
3. Management of resources. To manage the continuity of resources to guarantee the academic and administrative processes, based on the requirements of the educative project.	*a)* To plan the handling of the budget and the investments, according the institutional policy and the current norms. *b)* To control the budget execution according to the current, legal and the institutional policy. *c)* To obtain resources through the presentation of projects to different entities and organizations, according to the needs and opportunities of the educative institution. *d)* To manage contracts according to the institutional policy and the current norms.	*a)* Report of the financial management done in the organization. *b)* Report of the achieved resources and their impact in the academic and research processes. *c)* Report of the situation and maintenance of the resources of the institution, with their corresponding process of optimization.

Essential competences of an educative principal.	*Some fundamental criteria.*	*Some key evidences.*
	e) To evaluate the situation, use, optimization and maintenance of the resources or the institution in accordance with the needs of the educative activities. *f)* To show honesty in the handling of resources of the educative institution and to assure their care.	
4. Addressing of the certification process. To address the certification or accreditation process of the educative service, to assure the integral formation and the development of competencies in the students, teachers, the administrative personnel and principals with transparency, responsibility, with the established guidelines of the institution.	*a)* To coordinate the system of quality assurance of the educative service, based on the philosophy and the adopted guidelines by the institution. *b)* To orient the documents of the processes to certify or to accredit, according to adopted parameters by the institution. *c)* To evaluate the system of quality assurance of the educative service, based on the established criteria. *d)* To have the spirit of challenge to assure the quality of the academic and administrative processes of the institution.	*a)* Report of the process of the quality assurance in the institution with the corresponding documents. *b)* Report of the impact of the process of quality assurance in the institution.
5. Management of the human talent. To manage the talent of the personal under his	*a)* To determine the profiles by competencies of the job positions, in keeping with the institutional educative project. *b)* To coordinate recruitment, selection, and hiring of personnel, in accordance with the	*a)* Report with the management planning of the human talent in all its processes. *b)* Report of the execution of the different processes of the management of human talent per a year, with the evaluation of its impact.

command, with the objective of fulfilling with the purposes of formation and the development of competencies in the students, in accordance with the current norms, the strategic planning of the institution and the transparence of the actions.	established competency profiles and the current norms. *c)* To direct the formation and evaluation of the human talent, in accordance with the profiles of the job positions by competencies and the goals of the organization. *d)* To value the performance of the teachers in their classes or sessions and to assure the improvement in accordance with the established goals. *e)* To create the conditions which favor the organizational atmosphere, based on the requirements of the established projects. *f)* To direct the promotion and compensation (or incentives) of the institutions´ members in keeping with their performance. *g)* To direct the administration of the human talent with ethical commitment.	
6. Marketing. To address the offer of services and educative products to the community, so that interested people and with the expected graduation profile enter the institution, according to the educative offer, the requirements of the customers and legal norms.	*a)* To define the market niche , based on the characteristics and needs of the users. *b)* To formulate promotion and offer strategies of the educative services, depending on the target population. *c)* To coordinate the process of marketing in the educative institution, with honesty and social responsibility, with respect to the ethical norms which exist to achieve the goals.	· Report of the marketing process per a year, with evidence of the impact.

2. *Selection.* It consists in the set of activities which help to identify the institutional requirements of human talent. The possible candidates are sought and each candidate or candidates who fulfill with the job position profile is selected. To achieve this, it is recommended:

a. To ask for a portfolio of evidences which show they have the required competencies for the job.
b. To do an interview to have major information about the candidate´s competencies.
c. To do a written test about the conceptual and technical aspects.
d. To value the psychosocial aspects to evaluate generic competencies, such as team work, leadership, etc.

These evidences are evaluated based on a map to determine the level of achievement reached and to make a decision about the possibility of connection with the institution. This process finishes when there is a contract and the compromises are written of each one of the parts, which gives place to induction.

The selection of a teacher for each training space requires the evaluation by competencies, with the objective of assuring the minimum required suitability to mediate the learning, teaching and evaluation process in the students. This must be documented and to assure it to be with the expected quality.

3. *Evaluation.* This is a set of systematic, established activities, which help to evaluate to teachers, principals and administrative personnel regarding their competencies, and the reached results during a determined period. To achieve this, the base of the competencies profile is necessary and the plan of action of the period, with its corresponding goals. The evaluation is based in the following instruments:

a. Evidence- portfolio
b. Surveys to colleagues, chiefs, and students; according to the institutional policies which had been established about the topic.
c. Sometime, to do tests to evaluate conceptual and methodological aspects.
d. It is required sometimes to apply interviews to the person.

As it is in the selection, for the evaluation there is the use of an evaluation of competencies map . This map is similar to the one done with the students. With it, the different, established evidences in the jobs profile are valued, as well as the level of achieved performance. This report is given to the immediate boss so that they make the corresponding decisions, according to the institutional policies which exist.

4. *Formation.* The human talent formation is based on the evaluation of the principal formative needs of the teachers, principals and administrative personnel. For this, the profiles of the position jobs are taken into account, the reports of the selection processes and the evaluation reports. The formation is recommended to be through projects, with the objective of having impact on the teaching change, and administrative practices.

5. *Promotion and ending of the contract.* It consists on the set of activities and policies oriented for teachers, principals and administrative personnel to be promoted from one position to another one. It might be in the same area or in another one. It implies:

- The step from one position to another one in the same area and in the same level of complexity.
- The step from one position to another one in the same area but with a higher level of complexity.
- The step from one position of another area in the same level of complexity.
- The step of a position of other areas with a higher level of complexity.
- Ending of the contract and process of leaving the organization.

6. *Remuneration.* This is the process which helps to determine the pay to the teachers, administrative personnel and principals, by following the legal norms and by considering the performance evaluation in each period (for example, each six months or each year). The pay depends on the complexity of the competencies of the job as well as the achievement of goals.

Actions:

a. To determine the base salary of the teachers, administrative personnel and principals according to the nature of each job and the complexity of the required competencies.
b. To establish a variable, complementary salary or salary compensations, according to the achievement of goals of each position and the performance during a determined period (six months or a year).

7. *Quality of life.* It consists in the process which helps to assure that the teachers, principals and administrative personnel have the necessary resources to integral health, social security and recreation. Some actions are:

To evaluate-to communicate.

1. To determine if the profiles of the teachers, principals and administrative personnel are established with base on the competencies and they correspond with the institutional, educative model, the legal norms, and the challenges of the context.

2. To evaluate if the selection, evaluation, formation promotion, remuneration, and quality of life processes of the teachers, principals, administrative personnel agree with the legal norms, the institutional, educative model, and the challenges of the context.
3. To document all the action of planning and execution of the process, as well as their evaluation, by having in mind the documents of the meetings of making decisions and the processes of socialization and of argument.
4. To socialize with the educative community the processes of selection, evaluation, formation, promotion, remuneration and management of quality of life of the teachers, principals, administrative personnel, and to achieve basic agreements for their implementation.

Process 11. Management of resources

Concept. *The resources* are the media or supplies which help to do the activities of the different curricular processes, by considering the addressing, planning, performance and evaluation-communication. In general manner, the resources are divided in the following components:

- Physical spaces.
- Equipment
- Materials
- Bibliography
- Money resources.

In the tackling of the resources it is precise to have into account the possible need of adapting them or innovating them according to the needs of curricular management. It is also precise to establish actions to give maintenance to determined resources, to assure their pertinence and usefulness.

To address: product: document with the description of the essential resources to implement the formative process and with the policies for the management of the resources.

Criteria:

1. The established resources agree with requirements of the different processes of curricular management.
2. The key actions are identified to adapt, to create or to innovate the resources, according to the requirements of the processes of curricular management.
3. A mechanism, to plan and to register the use of the resources in the different processes of curricular management, are determined.

To plan:

1. To review and to adjust the timetable of planning of the resources to implement the program of studies by competencies.
2. To determine how to identify the necessary resources in the different curricular processes.
3. To determine the process of participation of the educative community in the identification of the necessary resources in order to implement the program of the study of competencies.
4. To establish the necessary actions to achieve the required resources in the curriculum by competencies.

To act:

1. To review each one of the processes of curricular management and to determine the necessary resources for its implementation. To consider the established criteria in the addressing.
2. To look for participation of the educative community in the identification and following up of the resources to implement the curriculum by competencies.
3. To establish the necessary actions to get the resources.
4. To determine the process of maintenance of the resources.
5. To establish the process of evaluation of the quality of the resources, by looking for their validity, pertinence and usefulness.

To evaluate-to communicate

1. To establish if the resources agree with the needs of the processes of curricular management.
2. To evaluate the actions of maintenance and adaptation of the resources.
3. To determine if the actions of searching of resources are feasible, according to the opportunities of the context and the institutional possibilities.
4. To document the process of identification of resources, as well as the actions of searching, adaptation, and assurance of the quality of these ones.
5. To socialize the plan of resources in order to do reality the implementation of the curriculum by competencies.

Process 12. Mediation of the integral formation.

Concept: *The mediation* is the center of all curricular process. It consists on achieving that the students have an effective integral formation and the necessary competencies to face the challenges of the different contexts, as in the present as in the future, in a local, national and global plan. This implies to implement concrete actions in each class or learning session to contribute to the achievement of this purpose, by considering the entry and graduation profiles, the curricular map, and the planning of each training space.

In the frame 3.20 are described the differences of mediation with other traditional processes, such as the teaching and easement.

The mediation is the most integral process because it implies to give learning opportunities to the student , but he is also demanded commitment, effort and perseverance. It does not mean that his learning must be directed continuously, as it happens in traditional teaching, it does not try to make his learning much more simple or to help the "easement", but to achieve him to have a more active role, according to a clear working plan and with certain norms of study, with the help and assessory of a competent teacher who knows his role clearly , he knows he is not an social animator, nor a father or a friend (as it many times happens in the teachers who assume themselves as "facilitators").

Frame 3.20 differences of mediation with the teaching and easement.

Aspect	Teaching	Facilitation	Mediation
Approach	It is centered in the role of the teacher, who plans, executes, and evaluates the learning process in the students, according to determined frames of reference.	It is centered in the students, in their rhythms and learning styles, by considering their needs and interests. The teacher orients the process by considering the students.	It also considers the rhythms and learning styles, but besides, it is also looks for the students and the teachers to work for the achievement of concrete goals with a sense of challenge and discipline.
Goal	It achieves the learning of contents and the acquisition of objectives.	The students learn competences, with the necessary resources, in an easy and fun manner.	The integral formation and the development of competencies as in the teachers an in the students, by tackling the challenges of the context, with dedication and perseverance. More than fun, it is searched to have formative classes.
Knowledge	It is centered in knowledge in a more cognitive form, although it also considers the other knowledge, according to the educative model which it has.	It tackles different knowledge: to know how to be, to know how to do, to know how to know and to know how to coexist.	It is focused in the students to develop and to set into action in an articulated form their knowledge, around the tackling of problems of the context.

Opportunities for the students to learn	It gives opportunities so that the students achieve goals in occasional form, and without having as a base the continuous improvement.	Multiple opportunities are given so that the students learn. The teacher assumes his task as to facilitate this with all the means as possible.	The evaluation is carried out with a formative goal and opportunities are given to the achievement of goals with base in this evaluation. Actions of help are implemented from the same students, the society ,the educative institution and the teachers.
Set into action	Basically, it is through master classes, and it is sometimes with classes of the students.	Application of multiple didactic strategies, looking for most of the time, the students to be motivated and happy with their study.	Application of the key 10 actions among the teachers and the students (frame 3.21) 1. increasing sensitivity. 2. Conceptualization 3. solution of problems of the context. 4. values and ethical project of life. 5 collaboration 6. Assertive communication. 7. Creativity and innovation. 8. Transversality. 9. Management of resources. 10. Evaluation-metacognition.

Action of mediation	Some key activities of the students
Opportunities for the students to learn	a) To visualize the competencies to be achieved and their relevance in the context. b) To set out and to answer questions about the contents and learning activities. c) To identify and to control distracting factors in the external context and to a personal level. d) To look for positive aspects in situations that show to be negative to reinforce the motivation.
2. Conceptualization	a) To apply conceptual maps to organize the ideas around a topic or activity. b) To set into action the conceptual cartography to dominate concepts and relevant theories. c) To look for, to process, to create, to criticize, and to apply the knowledge in the frame of a contextual problem.
3. Solution of problems of the context.	a) To apply the earning based in problems with their different phases. b) To apply the heuristicV to identify to interpret, to argue and to solve problems of the context.
4. Values and ethical project of life.	a) To apply the method of moral problems to learn how to discern of how to act with base on the universal values. b) To do live exercises which imply to set into action the universal values. c) To reflect upon the application of the universal values with base on a personal dairy.
5 Collaboration	a) To do cooperative work to achieve a determined goal, with a definite time and concrete resources. b) To execute socio dramas which deal with challenges of the context.
6. Assertive communication.	a) To do role plays to learn how to have assertive communication in different situations of social challenge. b) To analyze videos with situations of aggressive communication, non assertive and assertive, to understand the differences. c) To register the communicative situations everyday to learn how to regulate them.

7.Creativity and innovation.	a) To apply mental maps in the generation of ideas. b) To create ideas through the handling of concrete objects, such as plastiline. c) To simulate innovations with base on challenges of the context.
8.Transversality.	a) To solve problems from the pluri diversity b) To solve problems from the interdisciplinariety, c) To solve problems from the transdisciplinariety
9.Management of resources.	a) To diagnose the resources through an action plan. b) To adapt resources based on the analysis, practice and error. c) To innovate resources based on the visualization
10.Evaluation-metacognition	a) To apply the MADFA strategy to improve in a continuos manner b) To apply self verbalizations to self regulate the performance. c) To implement the auditory to a mate to contribute to his continuous improvement.

To address. Product: portfolio with evidences that the teachers are forming the expected competencies in the students, by considering as minimum, the ten essential actions in the competencies with the metacognition (DAEC_M) (Tobón, 2010a). The evidences can be: videos, audios, certificates of the students, certificates of colleagues, evidences of evaluation of the same students, etcetera.

Criteria:

1. The mediation of the integral formation and the development of the competencies is taken to reality by taking into account the institutional educative model.
2. The process of mediation is implemented in the classroom based on the concrete activities. This achieves the active participation of the students.
3. The mediation is evaluated through evidences of performance with the students. This considers their achievements in the integral formation and the development of the expected competencies in the curriculum.
4. The actions of formation of the teachers have impact in the change of the traditional educative practices, in accordance to the goals of the institutional educative model.

To plan:

1. To review and to adjust the timetable of planning of the mediation actions since the leadership of the teachers.
2. To determine how the practices of the teachers are with the students and their impact in the achievement of the graduation profile.
3. To look for the necessary referents to establish the actions of mediation , which must be analyzed, argued, and agreed with the principals, teachers and students.
4. 4. To invite to the principals, teachers and students (together with other actors of the educative process) to give proposals around the key actions of mediation to assure the integral formation and the necessary competencies in the students.
5. To establish mechanisms to agree the actions of mediation to implement in the educative institution.

To act:

1. To diagnose the practices of the teachers and to look for them to establish actions of improvement, with provable commitment. With this goal, they are given in the portfolio of evidences and the evaluation of performance in the classroom.
2. To determine the key activities of mediation which the teachers

must implement with their students in the classroom.
3. To form to the teachers in concrete actions of mediation with the students, through socio dramas, and theoretical-practical activities.
4. To establish actions of help for the teachers so that they become mediators.
5. To evaluate the performance of the teachers by their role in the learning of the competencies through concrete evidences.

To evaluate-to communicate:

1. To establish if the actions of mediation implemented are coherent with the institutional educative model.
2. To determine if the actions of mediation involve to the students by looking for their active participation as constructors of their own learning.
3. To identify if the actions of mediation are feasible of being implemented by considering the institutional context and the external context, in such a form that it does not remain in paper.
4. To document the process of determination of the actions of mediation with the students in the institutional context.
5. To socialize the essential plan of mediation and to achieve the minimum agreements which make possible its implementation in a continuous manner and they change into institutional culture.

Bibliographic references

Argudín, Y. (2008). *Educación basada en competencias. Nociones y antecedentes [Education based on competencies. Notions and precedents]*. México D.F.: Trillas.

Deming, W.E. (1989). *Calidad, productividad y competitividad. La salida de la crisis [Quality, productivity and competitivity. The exit of the crisis]*. Madrid: Ediciones Díaz de Santos (edición original en inglés, 1982).

Elliot, J. (1994). *La investigación - acción en educación [The investigation-action in education]*. Madrid: Morata.

García Fraile, J.A., López Calva, J.M., López Rodríguez, N.M. y Aguilar Álvarez, A. (2012). *Gestión curricular por competencias en la educación media y superior. La perspectiva humanística del enfoque socioformativo [Curricular management by competencies in secondary and higher education. The humanistic perspective of the socio-formative approach]*. México: GAFRA.

García Fraile, J.A., y Tobón, S. (2009). *Estrategias didácticas para formar competencias [Didactic strategies to create competencies]*. Lima: A.B. Representaciones Generales.

Lev Vygotsky, L. S. (1985). *Pensamiento y Lenguaje [Thought and Language]*. Buenos Aires: Pléyade.

Martínez, S., Frade, L., Pimienta, J. y Tobón, S. (2011). *Evaluación de las competencias en pruebas masivas [Evaluation of competencies in massive tests]*. México: SEP y CREFAL. En prensa.

Monereo, C. (2007). *Estrategias de enseñanza y aprendizaje [Strategies of teaching and learning]*. Barceona: Graó.

Morin, E. (2002). *Los siete saberes necesarios para la educación del futuro [The seven knowledges required for the future education]*. Buenos Aires: Nueva Visión.

Perrenoud, P. (2004a). *Diez nuevas competencias para enseñar [Ten new competencies for teaching]*. Barcelona: Graó.

Perrenoud, P. (2004b). *Desarrollar la práctica reflexiva en el oficio de enseñar [Developing the reflective practice in the teaching office]*. Barcelona: Graó.

Pimienta, J. (2012a). Las competencias en la docencia universitaria. Preguntas frecuentes *[Competencies in university teaching. Frequent questions]*. Naucalpan, Estado de México: Pearson – Prentice Hall.

Pimienta, J. (2012b). Estrategias de enseñanza aprendizaje. Docencia universitaria basada en competencias *[Strategies of teaching-learning. University teaching based on competencies]*. México: Pearson.

Ramírez, M. y Rocha, M.P. (2007). *Guía para el desarrollo de competencias docentes [Guide for the development of teaching competencies]*. México D.F.: Trillas.

Schon, D. (1992). *La formación de profesionales reflexivos: Hacia un nuevo diseño de la enseñanza y el aprendizaje de las profesiones [The formation of reflective professionals: Towards a new design of teaching and learning of the professions]*. Barcelona: Paidós.

Schon, D. (1998). *El profesional reflexivo [The reflective professional]*. Barcelona: Paidós.

Tobón, S. (2001). *Aprender a emprender. Un enfoque curricular [Learning to undertake. A curricular approach]*. Medellín: Funorie.

Tobón, S. (2002). *Modelo pedagógico basado en competencias [Pedagogical model based on competencies]*. Medellín: Funorie.

Tobón, S. (2004a). *Cartografía conceptual [Conteptual cartography]*. Islas Baleares (España): CIBER EDUCA.

Tobón, S. (2004b). *Formación basada en competencias. Pensamiento complejo, diseño*

curricular y didáctica(1era. Ed.) [[Formation based in competencies. Complex thought, curricular design and didactic (1st Ed.]. Bogotá: ECOE.
Tobón, S. (2006). *Las competencias en la educación superior. Políticas hacia la calidad [Competencies in higher education. Policies towards quality]*. Bogotá: Ecoe.
Tobón, S. (2009a). *La formación humana integral desde el proyecto ético de vida y el enfoque de las competencias [Integral human formation from the ethical project of life and the competencies approach]*. En E. J. Cabrera (Ed.), *Las competencias en educación básica: un cambio hacia la reforma [The competencies in basic education: a change towards reform]*. México: Secretaría de Educación Pública.
Tobón, S. (2009b). *Proyectos formativos: didáctica y evaluación de competencias [Formative projects: didactics and evaluation of competences]*. En E. J. Cabrera (Ed.), *Las competencias en educación básica: un cambio hacia la reforma [The competencies in basic education: a change towards reform]*. México: Secretaría de Educación Pública.
Tobón, S. (2010a). *Formación basada en competencias. Pensamiento complejo, diseño curricular y didáctica [Formation based on competencies. Complex thought, curricular design and didactics]*. Bogotá: ECOE.
TTobón, S. (2010). Proyectos formativos: metodología para el desarrollo y evaluación de competencias *[Formative projects: metodology for the development and evaluation of competencies]*. México: Book Mart.
Tobón, S. (2011a). *El currículo por competencias desde la socioformación. ¿Cómo podemos cambiar nuestras prácticas educativas para asegurar la formación integral? [The curriculum by competencies from the socio-formation. How can we change our educational practices to ensure integral formation?]* México: Limusa.
Tobón, S. (2011b). *La evaluación de las competencias por medio de mapas de aprendizaje: una propuesta frente a los métodos tradicionales de evaluación [The evaluation of competences through learning maps: a proposal against the traditional methods of evaluation]*. En Moya, J. y Luengo, F. (coords.), *Estrategias de cambio para mejorar el currículum escolar: Hacia una guía de desarrollo curricular de las competencias básicas [Strategies of change to improve the school curriculum: Towards a curriculum development guide of the basic competences]* (pp. 135-149). Madrid: Ministerio de Educación. Disponible en: http://www.cife.ws/comunidad
Tobón, S. (2011c). *Manual sintético de gestión del talento humano por competencias y la Quinta Disciplina [Synthetic manual of human talent management by competencies and the Fifth Discipline]*. México: CIFE.
Tobón, S. (2011d). *Evaluación de las competencias en la educación básica [Evaluation of competencies in basic education]*. México: Santillana.
Tobón, S. y García Fraile, J.A. (2006). *La formación de las competencias en la educación superior [Formation of competencies in higher education]*. Madrid: Universidad Complutense de Madrid.
Tobón, S. y Mucharraz, G. (2010). *¿Cómo aplicar las competencias en la práctica docente? [How to apply the competencies in the teaching practice?]*. México: Kundrivi.
Tobón, S. y otros (2006). *Manual de diseño curricular [Manual of curricular design]*. Bogotá: Dike.
Tobón, S., García Fraile, J.A., Carretero, M. y Rial, A. (2006). *Competencias, calidad y educación superior [Competencies, quality and higher education]*. Bogotá: Magisterio.
Tobón, S., Pimienta, J., y García Fraile, J.A. (2010). *Secuencias didácticas: aprendizaje y evaluación de competencias [Didactical sequences: learning and evaluation of competencies]*. México: Pearson.

Made in United States
Orlando, FL
05 November 2026

90230213R00133